# Confluence of Evolutionary Science and Christian Faith

# Confluence of Evolutionary Science and Christian Faith

## Toward an Integration

# Joseph Fortier

| Library of Congress Control Number: | | 2011918604 |
|---|---|---|
| ISBN: | Hardcover | 978-1-4653-8080-7 |
| | Softcover | 978-1-4653-8079-1 |
| | Ebook | 978-1-4653-8081-4 |

**To order additional copies of this book, contact:**
Xlibris Corporation
1-888-795-4274
www.Xlibris.com
Orders@Xlibris.com
104071

# CONTENTS

# ACKNOWLEDGMENTS

M ANY THANKS TO my students at Saint Louis University who took Evolution and Christian Theology for their thoughtful, insightful work that has inspired this book. Also many thanks to those faculty members who were so encouraging and supportive of the course and this book, especially Dan Finucane, Wayne Hellman, Ron Modras, Rob Phenix, Cornelia Horn, and Peter Bernhardt. Thanks also to the Jesuit community at Gonzaga University for their support and interest in this work. And many thanks to the people of the Blackfeet and Colville Indian Reservations who provided immense insight and motivation, without which this book would not have been written. Many thanks to my mother and father, who were my primary teachers concerning the things of the natural world and my faith.

# INTRODUCTION

S OME MODERN SOCIETIES in which both vibrant religious faith and scientific education thrive are plagued by an emotionally charged, combative attitude between those who consider themselves advocates of one side or the other. To wit, the spate of hostility by a few scientifically oriented authors who purport to address the topic of science and religion by assaulting religion, such as Daniel Dennett (*Darwin's Dangerous Idea*), Richard Dawkins (*The God Delusion*), and Sam Harris (*The End of Faith*). On the other hand, one only needs to surf the web for a short time on topics such as "evolution" and "age of the earth" to find pages full of half-truths by purportedly religious conservatives and others who rigidly hold to scientifically unsupportable alternatives to cosmic and biological evolution, such as Conservipedia's page "Evolution," and "The Age of the Earth." Are religion and science really at odds? Can one maintain one's intellectual integrity and maintain religious belief and spirituality, or has this become impossible? Do the popular books cited above really spell doom to faith in God by thinking people, or do they perhaps originate from a somewhat rigid, imagination-compromised mentality similar to the websites also cited above? This author maintains the latter and will explain why in this book.

I maintain that the underlying cause of the disjunction between science and faith in the worldview of many people is a sort of complacent lack of awareness on the part of the authors they read. The scientific atheist authors cited above base their somewhat banal assaults against religion on words of some Christian authors and leaders that are neither mainstream nor historically aware, in contrast to the mountain of rigorous historical theology out there. These authors seemingly refuse to read outside their own comfort zones. Their attempts to attack and ridicule Christian faith and thought are without relevance because they don't know or understand what they attempt to attack and ridicule, so they go after what's easy: a sort of ahistorical biblical literalism that, while vocal, is certainly neither informed nor mainstream with respect to Christian thought. A major objective of this book is to represent historically enlightened, current Christian thought of which the above authors seem unaware.

Unfortunately, the positions by some Christian writers and leaders that I describe as ahistorical are those that, like the positions of the authors described above, show a sort of complacent lack of awareness about areas of science that they find challenging and uncomfortable, such as evolutionary biology. They systematically refuse to understand the science they attack. Like the above authors, they seem not to read or attempt to understand outside their own comfort zone, refusing to accept the reality of overwhelming scientific discovery in many areas of inquiry—including geology, biochemistry, morphology, anatomy, physiology, cell biology, embryology, genetics, and physics—that points with powerful and elegant congruence to the fact that cosmic and biological evolution has happened and is happening.

This book is based on years of teaching an interdisciplinary course called Evolution and Christian Theology. The course was designed for undergraduates. Its goal was to help students understand that Christian faith and evolution are not discordant, and not enemies. One needn't take sides. I met with varying levels of success in the class. There was the deeply religious student from a conservative evangelical background who took the course to heart, seemed to accept the veracity of scientific discovery, but then near the last teaching day, stood up and repudiated the entire corpus of the human intellectual heritage that did not agree with his tradition's particular interpretation of biblical scripture—including, of course, biological evolution. There was also the young atheist whose final paper for the course is the finest example of lyrically beautiful and rigorously thoughtful integration of Christian thought with biological evolution that I read in my four years of teaching the course.

In this book, I have attempted to chart a path for the reader toward understanding how rigorous empirical scientific thought and solid, informed as well as inspired religious faith are in fact congruent. The instruments that are used in this book to chart this path are as follows: (1) the findings of scripture scholars regarding the Genesis creation narratives; (2) the basic biology of evolution and genetics; (3) the philosophy and theology behind the positions people develop with respect to evolution and religious faith; (4) the basics of the science of complexity, which is at the confluence of science and religious faith; (5) the evolutionary thought of Pierre Teilhard de Chardin, which also lies at the confluence; and (6) the evolutionary theology of nature of John Haught, which offers the panoramic viewpoint at which one might glimpse the confluence.

Out in the forest, there are two clear-water streams that, at close range, apparently flow from different mountaintops but which, when viewed from farther away, can be seen to flow from two ridges of the same mountain. Somewhere in the forest, the two streams join—become confluent. The purpose of this book is to find this confluence.

# CHAPTER 1

# Do The Genesis Creation Accounts
# Conflict with Scientific Evolutionary Theory?

WHY DO SO many people of religious faith feel that evolutionary theory, especially biological evolution, is incompatible with their faith in a creator god and in a universe in which there is purpose? Perhaps it is in part because of how they read the Genesis creation accounts in the Bible (Genesis 1:1 to 2:4a; Genesis 2:4b to 3:24). In this chapter, we'll address this issue by taking an in-depth look at these accounts to find whether there is a basis for this sense of their incompatibility with evolutionary theory. Questions will be posed that suggest major potential conflicts and then addressed.

## How can one find the literal meaning
## of a given biblical passage?

Does the Bible teach that Adam and Eve were two historical people who lived about four thousand years ago and are the biological parents of all subsequent humans in Genesis 2:4b to 3:24? The apparent answer to a modern reader assuming a contemporary meaning to the words in scripture is yes, it does. If you open a Bible to its very beginning (Genesis 1:1) and read through Genesis 2:4a, you will find that in fact it does say that the earth and living creatures, even the sun, moon, and stars, were all made by God in six days; and then God rested on the seventh. If you read from Genesis 2:4b to 3:24, you will come to the conclusion that the answer to the second part in the above question about Adam and Eve is yes. It is exactly this reading that causes conflict between some very well-intentioned people of Christian faith and equally well-intentioned scientists who look at all the geological, biochemical, and biological evidence and say, "The biblical account just does not square with the evidence." In fact, scientist or no scientist, this is not as easy a question to answer as it may first appear.

So is there a real conflict between the scientific evidence and what the Bible really says about the age of the earth and universe and the origins of our species? Does the Bible really assert that the earth and universe are ten thousand years old or less? Does it authoritatively disagree with the five hundred thousand times older age of about 5 billion years that scientists would have us accept for the age of the earth? To begin to answer this question, let's take a look at what the biblical scholars have to say about the Genesis creation accounts. But first, why give credence to biblical scholars? These are people devoted to uncovering the true, full meaning of the Bible in the cultural, historical, human context in which the inspired writers lived. These people have expertise in reading ancient languages that were used during the time the Bible was written and can compare biblical passages with other ancient literature. When they do so, they often find compelling evidence that a biblical author's meaning is most accurately discovered in the author's historical, cultural, and human context—the context in which God works with us and inspires us.

Biblical scholars refer to the intention of a given biblical author as the literal sense. This way of understanding the Bible takes the words at face value but also as they were intended by the original author(s) or editor(s), rather than as they have come to be understood in today's context (Bergant 1989, 27) There is a difference between this literalist approach and another literalist approach sometimes used today in which the reader assumes a contemporary meaning to the words. The problem with this second approach is that the reader assumes that today's world of meaning is identical to that of the writers of a given biblical passage (Bergant 1989, 27). As we shall see, this assumption is often not correct.

What have the biblical scholars found about the literal meaning of the Genesis creation accounts in Genesis 1:1 to 2:4a and Genesis 2:4b to 3:24? To answer this question, let's take these accounts one at a time, beginning with the first one. We'll explore for answers to the following questions:

1. Who wrote Genesis 1:1 to 2:4a?
2. What cultural factors affected the author's worldviews?
3. What historical events affected the author's worldviews?
4. What sort of literature is Genesis 1:1 to 2:4a?
5. What major religious issues does this account address, especially in light of questions 2 and 3 above?

*Who wrote Genesis 1:1 to 2:4a?*

The consensus among biblical scholars is that the first creation account (Genesis 1:1 to 2:4a) was written by an author around the time of the

Babylonian captivity of Israel, around or after 587 BC. Biblical scholars call this author the priestly author. Evidence shows that this author belonged to a long scholarly tradition among the ancient Hebrews and summarized the earlier writings of this tradition. His work is included in several places in the first four books of the Bible (the Tetrateuch) besides the first account of creation (Bergant 1989, 35–36; Speiser 1964, xxvi).

When Genesis 1:1 to 2:4a, the six days of creation, is compared with Genesis 2:4b to 3:24, the story of Adam and Eve, biblical scholars have noted distinctive writing styles and themes that are also found in other passages in the Tetrateuch. In comparing Genesis 1:1 to 2:4a with Genesis 2:4b to 3:24 (the second creation narrative), the first creation narrative is rigidly structured and repetitive, giving it a formal, regal tone. The writer refers to God with the formal *Elohim*, or Lord God, and avoids portraying God in humanlike fashion (anthropomorphism). Chapter 2 verse 4 hints that the writer is concerned with genealogy of creation. This concern for genealogy is found throughout the Tetrateuch, wherever this unique formal writing style with use of *Elohim* for God is found. For example, see Genesis chapter 5.

The writer of Genesis 2:5 to 3:24 is known as the Yahwist because only this author refers to God as *Yahweh* in all his writings in the Tetrateuch. Unlike the priestly author, the Yahwist's writings here and throughout the Tetrateuch show a colorful folktale style and concern for the character development of his human subjects. He anthropomorphizes God, portraying God as conversing with humans and otherwise feeling and behaving in humanlike ways. The Yahwist lived during the tenth century BC, during the age of Solomon and David, according to the consensus of biblical scholars. Thus, the creation narrative in Genesis 1:1 to 2:4a was written by a different author at a different historical moment than that of the creation narrative in Genesis 2:4b to 3:24.

*What cultural factors affected the author's worldview?*

When the Genesis first account of creation is compared with the Babylonian creation narrative, the *Enuma elish*, it is apparent that the priestly authors were influenced by the Babylonian work. The similarities between these works could only have been an influence by the Babylonian work on the priestly authors because the *Enuma elish* was written before the biblical narrative of the first creation. In fact, the completed form of the *Enuma elish* was written in the seventh century BC, about two hundred years before the biblical account was finalized. From earlier stories that appeared in the *Enuma elish*, especially concerning the Sumerian god Marduk, most biblical scholars agree that earlier versions of the account date to about 1800 BC. The document apparently deals with root ideas and literature of ancient Mesopotamia.

The following chart illustrates the similarity in details and sequence of events in the two creation narratives (adapted from Speiser 1964).

| *Enuma elish* | Genesis |
|---|---|
| Divine spirit and cosmic matter are and coexistent and coeternal. | Divine spirit creates matter and exists independently of it. |
| Primeval chaos. Ti'amat is cloaked in darkness. | The earth is a desolate waste. Darkness covers the abyss. |
| Light emanates from the gods. | Light is created. |
| The firmament is created. | The firmament is created. |
| Dry land is created. | Dry land is created. |
| Heavenly luminaries are created. | Heavenly luminaries are created. |
| Man is created. | Man is created. |
| The gods rest and celebrate. | God rests and makes holy the seventh day. |

The six days of creation and the gods' party afterward in the *Enuma elish* are written on seven clay tablets. In the Hebrew creation narrative, there are six days of creative activity and a day of rest.

Ancient Mesopotamian "science" was really a blend of religious ideas with notions of how the world and the heavens came to be, how they are constructed, and how they work. We saw above that the ancient

Mesopotamian notion of how the universe came to be (cosmogony) was quite religious. Another ancient Mesopotamian and Egyptian idea was that of how the earth and heavens are constructed. It was generally assumed that the earth is flat and that the sky is a solid dome thought to have been made out of metal, held above the earth by pillars (Gier 1987, chap. 13; Hasel 1972; Seeley 1991, 1997). There were thought to be windows or doors in the sky that open and close to let in precipitation—water, snow, and hail. This precipitation comes from the seawater above the sky-dome. Below and surrounding the earth is freshwater and seawater (Whybray 1995, 41). The celestial bodies are affixed in the solid sky-dome (Seeley 1991).

Bible scholars have observed that the Bible expresses the same cosmology (form and functioning of the world, universe) as these surrounding cultures, which really isn't surprising. The biblical authors weren't attempting to do science—they were assuming the cosmology generally held in their times and regions. As we shall see, they were much more interested in theological questions than cosmological ones. Thus, Genesis 1:6–8 describes the sky (*raqia*) as a solid substance that separates the water below and surrounding the earth from the water above the sky. In Genesis 1:14–19, the celestial bodies are placed *in* the sky—exactly as Mesopotamian and Egyptian literature has it. They are fixed in the solid sky-dome. Genesis 7:11 describes the beginning of the flood in the Noachian account in terms of waters below and waters above the sky. According to the Revised Standard Version of the Bible, "All the fountains of the great deep burst forth, and the windows of the heavens were opened."

In summary, the evidence is strong that the cosmology described in the biblical narrative of the first creation is also that of the surrounding Mesopotamian and Egyptian cultures. The identical sequence of the same events in the Genesis first creation narrative as those in the *Enuma elish*—the allusions to a solid sky-dome, celestial bodies fixed in the dome, water above the sky and water below the earth—the earlier date of the *Enuma elish*, and very early dates of accounts that were incorporated into it all provide strong evidence that the priestly authors drew from Mesopotamian culture.

*What historical events affected the author's worldview?*

Genesis 1:1 to 2:4a was completed around the time of the Babylonian captivity of the Israeli state of Judah, which began in 598 BC and culminated in 587, when Jerusalem was taken by the army of Nebuchadnezzar, king of Babylonia (Bergant 1989, 36; Bright 1981, 326–344; Speiser 1964, xxiv–xxvi). At a stroke, the state of Israel was destroyed, the old national-cultic community was broken, the temple was razed to the ground, and the national community left as a group of beaten individuals with no external evidence of being a people any longer. Thousands died of starvation or disease, many were executed, and many fled to save their lives. Little is known of what happened within the borders of Judah during the next fifty years. It is known that people returned on pilgrimages to the blackened site of the temple, which they recognized as a holy site. The wonder is that remnants of the people of the defunct state maintained their identity after the Babylonian invasion (Bright 1981, 326–344).

The intellectual cream of Jewish society was deported to Babylon. According to the biblical book of Jeremiah (Jeremiah 52:28–30), the total number deported was 4,600, which probably only accounts for adult males. These people managed to stow the records and written traditions of Israel up to that time. Through these documents, which recounted Yahweh's past deeds to his people and provided them with hope for the future, the community lived. Among these exiles were representatives of the priestly tradition who, referring to the old documents of their tradition, finalized a theological history of the known world, from creation to the commandments given to Moses at Mount Sinai (Bright 1981, 345–350). Among these writings was the first account of creation in what became the book of Genesis.

*What sort of literature is Genesis 1:1 to 2:4a?*

The first account of creation is not historical literature in the sense that historical events are recorded during the times in which they happen. The subject matter here is the creation of the universe before humanity even existed up through humanity's coming into existence. What we have here is a creation myth. It is important to understand the terms *myth* and *mythology* in their literary technical sense rather than in a modern rhetorical sense that means "something which isn't true." In its literary sense, a myth is a highly symbolic story devised by a group of people, using imagery and ideas they are familiar with, that gives form to deep truths not bound by time, beyond literal description, and beyond a person's ability to completely understand. Creation myths approach this level of truth through stories about how things came to be. The originators of these stories realized that they were not dealing with history, although at times those to whom the stories are passed on may come to consider the stories historical (Bergant 1989, 37; Speiser 1964, 27). Often though, the tellers of a story that comprises a creation myth do realize that they are using a given story to convey deeper truths and that the language they use is highly symbolic in a way that calls the reader/listener beyond the mundane. The first story of creation in Genesis is such a story. As we have seen, this story borrows heavily from more ancient Mesopotamian sources for its science; that is, its cosmogony (how the world came to be) and cosmology (the form and functioning of the world). The Mesopotamian story itself, with its science, is a cultural artifact that people were familiar with. The priestly authors who developed the Hebrew creation myth edited the Mesopotamian material in a way that communicates original Hebrew inspiration, as we shall find below.

*What major religious issues does this story address?*

As we saw, the final version of this creation myth was written during a dark time for the Jewish people, just after the Babylonians had utterly destroyed their nation and scattered, killed, or captured their remnants. The Jewish spiritual and political leaders were living in captivity in Babylon. They felt that the utmost need of their people was to preserve their relationship with Yahweh and keep alive the hope of his promise to them. The first story of creation became a potent tool for achieving just this. Elements of a story identified at that time with the Babylonians (*Enuma elish*), containing science and religious mythology, were thoroughly Hebraized. The story became a sort of spiritual declaration of independent Jewish identity.

How was this accomplished? The cosmology and cosmogony of the original Mesopotamian story was refitted with a Jewish theology in obvious contradistinction to the Babylonian one. The number seven takes on a particular Jewish meaning of wholeness. God's activity in the process of creation and his relationship with creation becomes Jewish. Finally, the meaning and importance of human existence takes on an importance and dignity peculiar to Jewish thinking.

In Jewish thought as well as Mesopotamian thought, the number seven symbolized wholeness, completion. In Jewish life, this was lived out in six days of work followed by a seventh day of rest (Leviticus 23:3). In ascribing this same sequence of work and rest to God's activity in creation, the priestly authors were asserting the godliness of the Jewish way of living and, in effect, appropriating the connection between wholeness and the number seven to the Jewish way of life and religious belief.

Jewish theology was unique in the Mesopotamian world in its monotheism. To ancient Jewish people, Yahweh was the one god. Interestingly, the six days of creation becomes a repetitive rhetorical device in which the major Jewish themes of the nature of God are accentuated. On each day, God creates by saying, "Let there be . . ." God remains distinct from his creation except as its creator. Unlike the Babylonian god Apsu, God does not consort with the sea monster (Tiamat) as a marriage partner or kill another god (Tiamat) as Marduk did and form the earth from her. Instead, God alone is the uncaused first cause of creation and existence, and all existence is dependent on God for its being. No aspect of creation is to be worshipped as a god.

Jewish theology is also unique in the moral value it confers on creation. Just as each "day" begins with "And God said let there be" or a close variation, each "day," except the second, ends with "And God saw that it was good." The latter phrase is found seven times altogether. The *Enuma elish* treats the creation of humanity as an accidental fallout from the cosmogonic battle among the gods. Humanity then becomes a slave to the gods. In stark contrast, the first creation narrative presents God as willing the creation of humanity as a special, privileged kind of creature. For example, in 1:26, "Then God said, I will make man in my image, after my likeness." In Egyptian belief, the pharaoh was created in the image and likeness of the god Ra—the ka of the god (Massey 2004, 133). Far from being an accident and a slave, man is specially created and given a privileged, pharaoh-like status by the one god, Yahweh.

The priestly creation account in Genesis was a Jewish manifesto of original inspiration. It is unique in the ancient Near East in its monotheism, its clear-cut sense of divine transcendence and the one god's role as creator of being, its ethics with respect to the primordial goodness of creation and also with respect to the dignity of humanity. The writers were not concerned with what was then the cosmogonic and cosmological consensus of the day; they assumed it. They were concerned with larger, timeless theological issues concerning the nature of God, creation, and humanity and the relationships among these entities. The seven major events of creation were borrowed from the *Enuma elish*, converted to days, and used as a rhetorical device to repeat these major theological positions and to affirm the Jewish way of life as consonant with the image and likeness of God.

JOSEPH FORTIER

It can be seen that the priestly creation narrative does not attempt to assert a scientific statement about the ages of the earth and universe. The number seven was widely used as a symbol of completeness in the ancient Near East. The *Enuma elish* divided the number seven into six events of creation and a day of celebration. It was written on seven clay tablets. The priestly authors used seven days, divided into six workdays and a day of rest. Since the number seven was thus used as a rhetorical device to repeat larger nonscientific theological truths and not to assert scientific fact, it can be seen that there is no conflict between the literal meaning of this passage and the findings of modern science.

Does the Bible teach that Adam and Eve were two historical people who lived about four thousand years ago and are the biological parents of all subsequent humans in Genesis 2:4b to 3:24?

Many of us have grown up in families and schools in which we have read and heard the very vivid, colorful account of Adam and Eve in the garden from Genesis 2:4b to 3:24. It is a comfort to feel that we know our first ancestors as the author of this passage introduces Adam and Eve to us. It is comfortable to feel that we now understand the historical reason for suffering, death, and evil in the world. Is this story also to be seen as mythology rather than history? Must we lose the comfort of knowing these larger-than-life people as historical figures?

In order to answer these questions, we need to turn again to what the Bible scholars have to say about these questions:

1. Who wrote this creation narrative?
2. What cultural factors affected the author's worldview?
3. What sort of literature is Genesis 2:4b to 3:24? Is it history or mythology?
4. What major religious issues does this story address?

*Who wrote this creation narrative?*

The story was written during the Solomonic empire, around 1000 BC, as part of a larger work by one of the most gifted figures in world literature (Bergant 1989, 36; Speiser 1964, xxvi–xxix). The writer is known as the *Yahwist* because God is referred to by name (Yahweh) only in his works. Other than the consistency in referring to God as Yahweh, there are few other distinctive words, phrases, or themes consistently and uniquely used by this author. What does distinguish the Yahwist is his brilliant writing style. It is clear and direct, with consummate simplicity, even as the plots and especially character development and motivation portray rich, vivid complexity. The inner lives of characters are revealed by their actions rather than through description; this is what attracts the reader's attention and fascination and draws him or her into the drama. The characters are so much themselves, so candid, that even their relationships with Yahweh are very human. Even God becomes anthropomorphic (Speiser 1964, xxvi–xxix).

It is not mere coincidence that while the priestly authors begin with "the creation of heaven and earth" (Genesis 1:1), the Yahwist begins with the "making of earth and heaven" (Genesis 2:4b). In the priestly account, center stage was given to heaven, and humanity was simply an item, however special, in a cosmic sequence of majestic acts. In contrast, in the Yahwist account, the earth is where the action is and humanity is at the center of interest. While the priestly tradition is heaven centered, the Yahwist is earth centered.

*What cultural factors affected the author's worldview?*

While the two traditions differ in writing style and in philosophy, the subject matter is basically the same in both versions. As with the priestly account, the Yahwist account derives detail from Mesopotamian traditions about first beginnings. For example, note the similarity in opening lines among the priestly authors, the Yahwist, and the *Enuma elish*:

Priestly author (Genesis 1:1):    "When God set about to create heaven and earth . . ."

Yahwist (Genesis 2:4b):    "At the time when God Yahweh made earth and heaven . . ."

*Enuma elish*:    "When on high heaven had not been named, firm ground below had not been called by name . . ."

Other evidence that the Yahwist borrows from Mesopotamian tradition is revealed in very old words that he uses for *mist* and for the name of the garden, *Eden*. Some background is in order at this point. There were three dominant civilizations in ancient Mesopotamia before and during the period in which the authors of the Tetrateuch lived. The

Sumerian civilization, beginning about 3500 BC, is the oldest literate human civilization. It began to decline around 2300 BC. The Akkadians, who conquered the Sumerians in 2350 BC, held Mesopotamia until the Amorites under Hammurabi took the region about 1728 BC. Babylonia adopted the written Akkadian language for official use but retained the Sumerian language for religious functions.

Now back to the Yahwist. Genesis 2:6 begins with "But a mist would well up from the ground . . ." The word the Yahwist uses for *mist, edu,* is an ancient Akkadian word. *Edu* in turn was derived from the Sumerian *a.de.a.* The word *Eden* itself is from the Akkadian *edinu* and the Sumerian *eden.* The Akkadian word is rare, unlike the Sumerian word. This fact testifies to the antiquity of the Eden tradition. The mention of the Tigris and Euphrates Rivers leaves no doubt about the location of Eden: Mesopotamia (Speiser 1964, 18–20).

The story of the Fall (Genesis 2:25 to 3:24) is fascinating for its allusions to old Mesopotamian stories and the genius of how it incorporates elements of these stories into an inspired Jewish spiritual treatise. Themes such as sexual awareness (e.g., nakedness), wisdom associated with loss of innocence, and nature's paradise are obviously not unique to the Bible. It is interesting though that all these themes are found in one passage in an old Sumerian document called the *Epic of Gilgamesh.* This story, which centers on the Sumerian hero-king Gilgamesh, was written over a period of 1,500 years and finalized about 2400 BC. It was written on twelve clay tablets. The passage of interest, the story of Enkidu, is on Tablet I (Dalley 2000, 39–47, 50–153).

Enkidu was created directly by the gods and was a wild man who lived with the wild animals away from human civilization, wore no clothes, and was described as a pure man. Enkidu was discovered by a woodsman's son while hunting. He reported to his father, who dispatched his son to report this sighting to Gilgamesh. Gilgamesh advised the son to take the temple prostitute Shamhat back with him to lure Enkidu away from the wild animals and tame him. This she did first with sex and then by teaching him to converse. After losing his virginity, wild animals fled from him, and he lost his running endurance. Shamhat told him, "You have become wise, Enkidu, you have become like a god." Then she gave him garments and taught him to eat cooked food and drink beer. He became "like any man" (Dalley 2000, 50–59).

It is almost certainly more than coincidental that this story shares so many motifs also found in the story of Adam and Eve: primeval innocence in a natural setting (the garden/wilderness), encounter with temptation leading to sexual awareness (including nakedness), wisdom, and loss of relationship with wild creatures (alienation from the garden).

Another story from the *Gilgamesh Epic*, "The Huluppu Tree," has the motif of a tree, a snake, and a woman. The snake is huge and dangerous. The woman's name is Lilith, and she lives in the Huluppu tree, as does the snake. She is referred to as the desolate maiden. In later Arabic folklore, she becomes a wisdom figure whose symbol is the owl. She is later associated with the Tree of Wisdom (Scerba 1999).

Whether the Yahwist had direct access to the *Gilgamesh Epic* is not known. Nonetheless, these stories were part of an oral tradition before they became written. These stories were probably widely known in the ancient Near East, especially since they originated from the cradle of that civilization: Mesopotamia. The presence of so many of their elements in the second creation narrative suggests that the Yahwist made use of traditions that originated in Mesopotamia, even though he well may not have been aware of that particular origin.

*What sort of literature is Genesis 2:5 to 3:24?*

The reader has probably guessed that this, also, is a mythical story rather than history. Like the priestly story, it uses the framework of a creation myth to deal with larger, timeless truths concerning the nature of God and existence. Also like the priestly story, the Yahwist's story teaches that God is the uncreated creator of the universe. Unlike the priestly story, the Yahwist portrays God as working with clay, breathing in it, relating with his work in an earthier, more intimate way. Similarly, God Yahweh relates with people in a human, earthy way, walking with them, talking with them, becoming disappointed with them, showing them compassion. The story deals with timeless existential issues such as the relationships between innocence, sexuality, and wisdom (culture, the world, self-promotion) and how this complicated set of relationships affects the human relationship with God and freedom. We see in this story that God Yahweh is compassionate and humble. After Adam and Eve fall to the allurement of the snake's temptation to eat from the tree of knowledge of good and evil, thereby losing their original innocence and relationship with Eden, Yahweh clothes them as Shamhat did Enkidu.

The name *Adam* represents a wordplay. Chapter 2 verse 5b reads, "And no man was there to till the soil." The Hebrew for *man* is *adam* and for *soil*, *dama*. In ancient near-Eastern literature, names were symbolic, not mere labels. They provided clues into the nature and essence of what was named. The association of these two closely similar words connotes the earthiness of the man, as though the name Adam means "earthling" (Speiser 1964, 16). There is a similar wordplay in naming Eve in 3:20: "The man [or Adam] named his wife Eve, because she was the mother of all the living." The Hebrew for *Eve* is *hawwa* and for *mother of all the living*, *hay* (Bergant 1989, 41; Speiser 1964, 23).

*What major religious issues does this story address?*

These were not names given to children by the people of the Yahwist's time or at any time, a fact that adds evidence to the timeless mythical nature of the story. Adam and Eve are everyman and everywoman (Bergant 1989, 41–42). Each of us experiences what Adam and Eve in this story experienced. We struggle with, negotiate for, and make compromises for our weaknesses, temptations, worldliness, and need for freedom, innocence, and God. Adam and Eve represent humanity's timeless experience of loss of innocence because of pride and need to negotiate a place in the world. As a consequence, we, like Adam and Eve, experience at once alienation from God and neediness of God's compassion and help.

Again we may find ourselves, in our wisdom, leaving the pleasing, innocent garden of believing that Adam and Eve were real historical personages. But in so doing, we may come to understand deeper, closer truths about ourselves and the dilemmas that each of us face in becoming experienced and versed in the ways of the world, and thus, we grow in compassion and respect for ourselves and one another. We may realize our urgent need for God's compassion as we negotiate our ways in life between the tug of our primordial innocence and the fact of dealing with the world and its moral ambiguity.

Again, we are not dealing with historical first people who lived about four thousand years ago. We are dealing with the ahistorical human condition. Again, there is no real conflict between the literal meaning of the Yahwist creation narrative and modern science.

# Summary

☐ In modern parlance, literalism is often taken to mean interpreting scripture in a way that assumes that the words meant the same at the time scripture was written as they do now. However, biblical scholars refer to the intention of a given biblical author's use of words as the literal sense rather than as they have come to be understood in the modern context. Since biblical scholars are extensively trained in ancient Near Eastern languages and culture, and for the purposes of brevity and clarity, this sense of biblical literalism is referred to in this book as informed biblical literalism while the sense of biblical literalism more commonly called by that name today is referred to as ahistorical literalism.

☐ The biblical passage Genesis 1:1 to 3:24 is composed of two creation narratives, one composed by the priestly author (1:1 to 2:4a) and the other by the Yahwist author (2:4b to 3:24). The priestly author is almost certainly a member of a long scholarly tradition who gathered and summarized that tradition around the time that the Babylonians captured Israel and exiled its leadership, around 587 BCE. This written tradition is characterized by a formal, strongly structured writing style, undeveloped characters in the narrative, and a preoccupation with the transcendent god (heaven centered).

☐ The Yahwist author lived and wrote during the Solomonic empire, during the peak of Israeli power and influence in the ancient Near East. The writing style of this author is colorful, with richly developed characters and plots, and with a preoccupation with the earth and world (earth centered). The Yahwist is considered by many literary critics to be one of the greatest writers to ever have set ink to paper.

☐ The evidence shows that the writers of the Genesis creation narratives borrowed their cosmology from more ancient Near Eastern sources. They were not scientists; scientists did not exist in the ancient Near East. They were inspired spiritual writers concerned with ultimate issues such as God and God's relationship with humans and the earth. Thus, such cosmological issues as the shape of the earth, where water was located, the amount of time it took for the heavens and the earth to come about, and whether humans came from two individuals or two thousand were not their issues. They were concerned about timeless truths such as the nature of God and God's relationship with humanity and the world, the goodness and corruptibility of humans, and why humans are the way they are.

☐ The number seven was a highly symbolic number for ancient Near Easterners. It symbolized fullness, completion, universality, just as it did during Jesus's time. As the Babylonians wrote their creation account on seven clay tablets, and as their six major creation events coincided in description and sequence with the six Jewish creation events in Genesis, so the seventh day in both cases was a day off. For the Jewish priestly writer of the first creation narrative, number seven symbolized completion and fulfillment. The six days of creation was used as a rhetorical device that allowed for repetition of the major theological themes: God calls creation into being by his word, and God sees that what comes into being is good. The six days also asserted the goodness and godliness of the Jewish way of life, in which work was conducted for six days and the Sabbath was a day of rest. This narrative does not conflict with the science of evolution since an informed literal interpretation does not interpret it as asserting a scientific fact about the amount of time it took for the universe and earth to emerge.

☐ In the second creation account, the names for Adam and Eve showed that they were not historical people but rather literary figures that symbolized everyman and everywoman. The story is not meant to be historical but rather to convey deeper timeless truths about the nature of people and their relationship with God. Thus, it was not intended to teach that all people who ever lived are descended from an ancestral population of only two.

# References

Bergant D. ed. 1989. Introduction to the Bible. In *The Collegeville Bible Commentary*. Collegeville, MN: The Liturgical Press.

Bright J. 1981. *A History of Israel*. Philadelphia: Westminster Press.

Collins F. S. 2006. *The Language of God: A Scientist Presents Evidence for Belief*. New York: Free Press.

Dalley S. 2000. *Myths from Mesopotamia: Creation, The Flood, Gilgamesh, and Others*. Oxford, Great Britain: Oxford University Press.

Gier N. F. 1987. *God, Reason, and the Evangelicals*. Lanham, MD: University Press of America.

Hasel G. F. 1972. "The Significance of the Cosmology in Genesis I in Relation to Ancient Near Eastern Parallels." *Andrews University Seminary Studies* 10: 1-20

Massey G. 2004. *Ancient Egypt—the Light of the World*. Sioux Falls SD: NuVision Publications.

Scerba A. 1999. "Gilgamesh and the Huluppu Tree (2000 BCE)." Last modified April 2009. http://www.geocities.com/Wellesley/Garden/4240/gilgamesh.html

Seeley P. H. 1991. "The Firmament and the Water Above." *The Westminster Theological journal* 53: 227-40

Seeley P. H. 1997. "The Geographical Meaning of 'Earth' and 'Seas' in Genesis 1:10." *The Westminster Theological Journal* 59: 231-55

Watson, J. D. and G. S. Stent. 1980. *The Double Helix: A Personal Account of the Discovery of the Structure of DNA*. New York: W. W. Norton & Company, Inc.

Whybray R. N. 1995. *Introduction to the Pentateuch*. Grand Rapids, MI: Wm. B. Eerdmans Publishing Company.

# CHAPTER 2

# What is Biological Evolution, and Does it Have a Mechanism?

F ROM THE DISTANT shadows of the past, people of all languages and cultures have speculated and told stories to attempt to answer the questions, Where did all this exuberant life come from? How did humans get here? In the West, many of us thought we had found the answers in Genesis. Then along came the Age of Enlightenment in the seventeenth century with its confidence in human reason and observation to answer our questions. The intellectual fruits of the Enlightenment have successfully challenged the cosmogony and cosmology found in Genesis on two fronts: modern biblical hermeneutics (analysis and interpretation) and empirical science. In chapter 1, we saw how biblical hermeneutics have identified much of the cosmogony and cosmology of Genesis with that of the entire ancient Near East and Egypt. In this chapter, we'll take a look at how modern science, born in the Enlightenment, actually works; what sort of questions are within its competency to address; and how it collaborates with biblical scholarship in addressing the cosmogony and cosmology of the ancient Near East, especially with respect to the coming into being of living organisms.

## How does science estimate the truth?

Now that we have looked at the evidence from archaeology, history, biblical theology, and the study of ancient Semitic languages and literature and, with that evidence, have found the grounds for a conflict with evolution based on the biblical creation accounts to be lacking, let's take a look at what biological evolution really is. In order to do this, we need to take a look at the underlying philosophical positions that science uses to discover things: *empiricism* and *philosophical realism*. Empiricism is a theory of knowledge that asserts that knowledge arises from the sensory experience of the evidence (Markie 2008). The correspondence of this sensory experience with what is really there is assumed by empiricism. Empiricism arises from philosophical realism, which asserts that reality is independent of our conceptual schemes. It holds that getting knowledge of the truth-value of a given concept is a process of continuing refinement in the correspondence of how the concept is expressed with observation of reality. Thus, mathematical realism holds that mathematical entities exist independently of the human mind. Humans don't invent mathematics; they discover it, as any other intelligent beings in the universe might, in a process of refining a concept so that it more accurately corresponds with what is observed. Again, the assumption is that reality really does correspond to our minds' and senses' perceptions of observed phenomena (Miller 2008). Thus, the scientific enterprise is a process of refining concepts by experimenting with and analyzing data accessible to the senses to test the accuracy (consistently observable correspondence with reality) and predictive power of scientific concepts.

*Hypotheses* (singular: *hypothesis*) are ad hoc concepts based on preliminary observation and experimentation. When repeated observation and experimentation within the scientific community consistently bear out a hypothesis in a way that causes skeptical scientists (skepticism is a major virtue in the scientific community) to agree that the evidence consistently supports it, the hypothesis becomes a generally accepted *theory*. Theories, however, remain open to testing, refinement, and the possibility of later rejection.

## What is biological evolution?

Evolutionary biologists like to explain evolution in terms of two broad ideas proposed by Charles Darwin, the nineteenth-century naturalist who first described evolutionary theory with empirical evidence for it. Those two ideas are *natural selection* and *descent with modification.*

Darwin got his idea of natural selection in part from the artificial selection that plant and animal breeders use (Darwin 1859). For example, if a maize breeder wants to produce a new strain of maize that yields more grain per acre of maize field, he only selects those plants in a field that produced the largest fruits, with most seeds per fruit. Only these does he allow to reproduce in order to grow the next seed crop (crop used for seed production). Maize plants with less than optimal fruits may be harvested but never allowed to pollinate and reproduce. The breeder may repeat this process through a number of generations until he has found maize plants that produce twice as much grain as the original generation did. He has selected *heritable traits* that result in maximal maize grain yield using the artificial method of culling out the less productive maize plants, presumably with heritable traits that do not result in maximal grain yield, and allowing only the most productive ones to breed with each other. He has artificially selected for maximal maize grain yield.

On his voyage around South America in the HMS *Beagle*, Darwin noticed in the Galapagos Islands that each of the three islands had a different kind of mockingbird. The mockingbirds were all similar to a kind that was found on the nearby mainland (Ecuador). He reasoned that a population of the mainland birds had migrated out to these volcanic islands or perhaps they were blown out in a storm. On each of the islands, a population had succeeded in surviving and reproducing, but birds had not migrated from one island to another. Since each island was somewhat unique from the others, each mockingbird population, over generations, had *adapted* to its particular island through *natural selection*. That is, those birds less able to cope with the island environment, and thus less able to live and reproduce, were culled out of the population by *natural environmental factors* while those best able to cope survived and reproduced. In short, the individuals with the heritable traits that conferred best adaptive value for the particular environment were selected for by natural, environmental factors. Darwin reasoned that after many generations, like a new cultivar of maize plant well adapted for grain yield, each island had a new *species* or genetically unique kind of mockingbird, best adapted to live and reproduce under the natural factors acting on it on its particular island. Darwin called this process *natural selection* because he saw it as similar to artificial selection by humans (Darwin 1859). Darwin recorded mountains of observations based on this South American trip that, along with these mockingbirds, all fell into a pattern that could only be explained by what he came to call *descent with modification* (Darwin 1859).

Descent with modification, Darwin reasoned, is found in both populations of domestic organisms undergoing artificial selection and populations of wild organisms undergoing natural selection. In both cases, the frequencies of heritable traits, that is, characteristics that are expressions of genetic information that is inherited from parents, change slightly in a population from one generation to the next. As a hypothetical example, imagine a population of rabbits in Canada before the first ice age, perhaps before it ever snowed in Canada. Most of the rabbits were brown, but occasionally a white one was born. Coat color is a heritable trait. The white ones didn't live long because there was a type of owl that loved to eat rabbits. The white ones were easy to spot and seldom lived to reproduce. As a result, the population consisted of mostly brown rabbits. Natural selection weeded out the white rabbits.

Along came the ice age—global cooling. As the winter snows lasted increasingly longer, the owls became increasingly fatter on the easy hunting since initially, it was easy to find the brown rabbits in the white snow. But the gene for white coat color was still present in the rabbit population, even though the *dominant gene* for brown coat color hid it if the white gene was also present in such a rabbit. The gene for white coat color was *recessive*. That is, the white gene is hidden unless each parent of a rabbit had a white gene along with its brown gene and both parents donated their white gene rather than any brown one to produce a white progeny. Lucky were the rabbits that inherited two white genes from such parents—now these white rabbits had the better chance of living to reproduce since they were less easily seen and caught by the rabbit-eating owls in the new snow-covered environment. Over time, the relative proportion of brown rabbits in the population gradually became less as the predators culled them, along with the gene in the population for brown coat color. White rabbits became more common and then predominant. The process of natural selection, the natural force of changing environmental conditions, had weeded out rabbits with genes coding for brown color from the population and had selected for white rabbit color. Over generations, rabbit color was modified (descent with modification) among these northern rabbits. Natural selection had pressured the rabbit population, over generations, to better adapt to snowy conditions.

The above explanation is an oversimplification, genetically speaking, but not far from the best explanation for the coat colors of present-day rabbit and hare species in North America. The arctic hare, which lives in northern Canada above the tree line, becomes white in the winter. In the short Arctic summertime, those in the southern part of the species's range turn brownish while those in the northern part of the range remain white. A relative of the arctic hare, the snowshoe hare, lives farther south in forested country up to the Arctic tree line. While these hares turn white in winter, they become darker brown in summer. Finally, the cottontail rabbit and jackrabbits, the ranges of which do not extend as far north as that of the snowshoe hare, remain brownish all year (Gray 2007). Of course, there are many rabbit and hare predators other than owls—various hawks, fox, lynx, and bobcat, among them.

# Gregor Mendel: is there a mechanism for biological evolution that maintains its integrity through generations?

Gregor Mendel, a Catholic priest, physicist, and horticulturalist who lived in a region of the Czech Republic in what was then part of Austria, was a contemporary of Charles Darwin's and admired his work (Henig 2000). He is known as the father of genetics. While Darwin published his treatise on evolution (*On the Origin of Species*) in 1859, Mendel published his on patterns of inheritance (*Experiments on Plant Hybridization*) as a scientific paper in the journal *Proceedings of the Natural History Society of Brunn* in 1866 (Henig 2000). Unfortunately, Mendel's work went unnoticed until 1900. If its importance had been recognized earlier and the connection had been made with Darwin's idea of descent with modification, the mechanism of evolution might have been discovered much earlier (Henig 2000). The lack of evidence for this mechanism of evolution, the elusive heritable factor, remained the Achilles' heel of Darwin's theory until Mendel's work was finally discovered (Henig 2000).

What was Mendel's research, what did he find, and how was it a significant piece of evidence for evolutionary theory? Let's begin with what he did. He raised pea plants—a lot of pea plants. He raised thousands of pea plants at any given time in the monastery garden where he lived. Garden peas were easy to grow, available in many varieties, and relatively easy to use for controlled pollination experiments. In flowering plants, which account for by far the majority of plant species alive today, the flower is the sexual organ (figure 2.1). As with many kinds of flowering plants, the pea flower has both female and male parts. The female part is the carpel. The male part is the stamen. In a pea flower, there is one pistil and ten stamens. The stamens produce pollen, tiny dust-sized granules, each of which carries sperm cells. When a pollen grain lands on the tip (stigma) of a carpel of the same plant species, the sperm is released, travels down the style, and finds its way to an ovule in an ovary at the base of the carpel. Inside the ovule is an egg cell, which the sperm cell joins with and fertilizes. A seed results, with an embryonic plantlet inside.

Sperm cell? Egg cell? What's a cell? A cell is the smallest unit of all living things. Some organisms, like bacteria and algae and their relatives, only have one cell. Others like daisy plants, squid, fungi, and us are composed of millions of cells. We have brain cells, nerve cells, skin cells, muscle cells, and many other kinds of cells that make up our bodies. The cells are tiny. Each one of us humans is composed of trillions of these tiny cells.

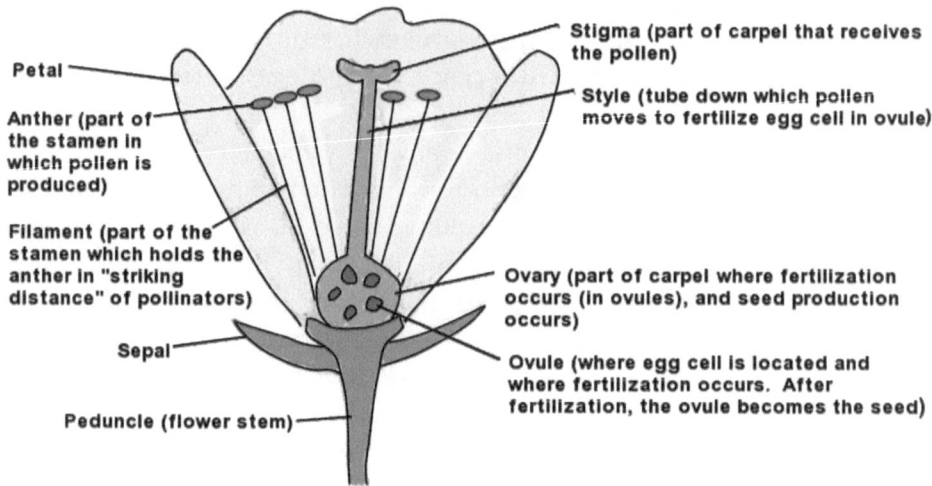

Figure 2.1. In plants that produce flowers, the flower is the reproductive unit. The carpels and stamens are, respectively, the female and male organs.

Unlike many plants, pea flowers can self-pollinate. If a bag is put over a flower to keep the pollen from the stamens of that flower from pollinating and fertilizing another flower, the flower will self-pollinate. Pea plants can be cross-fertilized, or sexually crossed, by transferring the pollen from the flower on a given plant with a fine brush to the stigma of a flower on another plant. The pollen from one plant will fertilize the flower of the other plant.

Mendel is credited with being a pioneer in the use of statistics to evaluate the results of experiments in empirical tests of scientific hypotheses (Henig 2000). From his large numbers of sexual crosses per experiment, he was able to calculate percentages of progeny showing various heritable traits with considerable precision. As we shall see, he also used a form of statistics, or number crunching, to discover and predict ratios of heritable traits resulting from crosses.

One of the first questions that Mendel asked was whether, in a sexual cross between two pea plant varieties, the heritable factors maintain their integrity or dissolve into one another. He raised "true breeding" varieties of pea plants with purple flowers and white flowers. Plants are true breeding for a given trait, such as flower color, if over a lot of generations, all progenies retain the same trait. When he crossed pea plants in which one parental true-breeding type had purple flowers and the other, white flowers, he found that their first generation progeny, the $F_1$ generation, all had purple flowers. No white flowers were to be found in this $F_1$ generation. He then crossed $F_1$

generation plants with other $F_1$ generation plants. Of the over one thousand progenies of the $F_1$ generation (the $F_2$ generation), 25% had white flowers and 75% had purple flowers. White flower color had reappeared. When he repeated this experiment several times, Mendel found that the results were consistent. He was able to conclude that not only does a heritable factor or gene exist but that the gene retains its integrity through generations. In other words, one variety, or *allele*, of gene doesn't dissolve into another. For the gene responsible for flower color in the peas Mendel raised, there was a version for purple color and one for white color. For a given gene, the *outwardly observable trait* expressed by that gene is called the *phenotype*.

But what happened to the alleles for white coloration in the $F_1$ generation? Why weren't they expressed? To answer this question, we must first look at two key features of the genetic systems of organisms.

One key feature of the inheritance system of most sexual organisms is that these systems are *diploid*. The word *diploid* is from the Greek *diplous*, meaning "double." *Diploid* means that there are *two copies* of any given gene in every cell of the organism. One copy is inherited from the female parent and the other from the male parent.

The other key feature is that of *alleles*. As we saw, each different *gene version* is called an *allele*. Thus, the male parent may contribute one allele (gene version), and the female parent may contribute another allele. Both parents would be contributing a copy of the same gene, just different versions of that gene. For example, perhaps my father contributed the brown allele for eye color. Perhaps my mother contributed the blue allele. Both parents contributed a gene for eye color—just different alleles (versions) of that gene. My phenotype then would be brown eye color. No blue to be seen. Why? The brown allele happens to be dominant over the blue allele for eye color in people. The blue allele is *recessive.* Dominant alleles *mask* the expression of recessive alleles. Or both parents may contribute the same allele. In that case, both parents would be contributing identical versions of the gene. The gene versions, or alleles, present in an individual are called the *genotype*. If each parent contributes a *different* allele for a given gene locus on a chromosome, the genotype is called *heterozygous*. If each parent contributes the *same* allele, then the genotype is called *homozygous*.

Why did the white color completely vanish between the parental generation and the first, or $F_1$ generation in Mendel's pea experiment, and then reappear in the second, or $F_2$ generation? To search for the answer to this question, Mendel applied a statistical technique called combination theory, which he picked up from his mentor, Andreas von Ettinghausen, a mathematician and physicist at the University of Vienna, who developed the theory. Using combination theory, Mendel charted the possible combinations of alleles of a given gene for a given pea plant generation and the probabilities of any given genotype (alleles present in an individual) and phenotype (outward expression of the alleles; for example, white or purple flowers) for the progeny of that generation. If $P$ symbolizes the allele for purple flower color and $p$ the allele for white flower color, then the *parental cross* could be symbolized by PP × pp since each parent came from true-breeding stock. *PP* would symbolize all alleles that are present at the locus of the gene for flower color in the parent with purple flowers. All alleles present for the gene for flower color in the parent with white flowers are represented by *pp*. As we saw above, in true-breeding purple-flowered pea stock, only purple-flowered plants ever result from sexual crosses between parental plants. The same goes for true-breeding white-flowered pea stock. Since pea plants are diploid, meaning there are two copies (maternal and paternal) of each gene present in every cell, in true-breeding pea plant stock, all plants have two copies of *only one allele* (homozygous genotype) of the gene for flower color: purple *or* white in Mendel's case. There are *never* two *different* alleles for a given trait (e.g., flower color) in plants that are true-breeding for that trait.

To most clearly explain Mendel's discovery, it's best to see his calculations through the eyes of the British biologist R. C. Punnett. In 1911, Punnett came up with a clear, simple way of charting Mendel's work. It is known today as the Punnett square (Henig 2000). The Punnett square for the parental cross of Mendel's pea experiment looks like this:

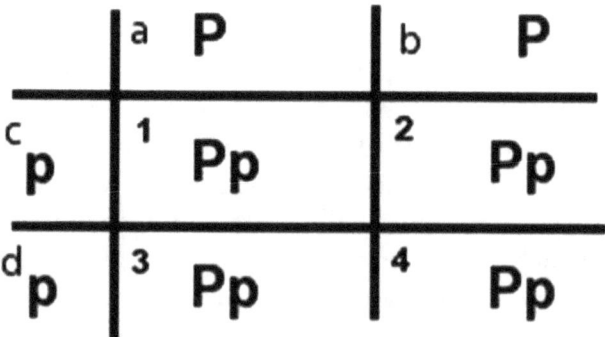

Figure 2.2. Punette square showing Mendel's parental cross, and all possible combinations of parental alleles on the progeny of that cross. First generation progeny (F1) of the parental generation are called the F1 generation. Parental alleles are labeled with letters. F1 progeny genotypes (alleles from both parents) are labeled with numbers.

For true-breeding individuals, the two genes for a given type of trait (e.g., flower color) are always the same version (same allele). In the square above, the possible gametes (sex cells) of the purple-flowered parent are aligned in the top row (squares A and B). The possible gametes donated by the white-flowered parent are aligned in the left column (squares C and D). The genotypes for the progenies for all possible combinations of gametes from each parent are shown in boxes 1–4. Each box (1–4) shows a possible combination of an allele from one parent and an allele from the other. Each box symbolizes a possible fertilization (egg + sperm) event. Thus, each box (1–4) represents a possible first generation ($F_1$) progeny genotype as well as the probability (1/4, or one out of four) that a progeny with this genotype will result from any given reproductive cross. All four boxes represent all possible allele combinations (genotypes) of progeny for this parental cross. For example, the alleles of the progeny represented by box 3 include the allele A of the purple-flowered parent and the allele D of the white-flowered parent. The alleles that the progeny represented by box 2 will inherit will be those from the white-flowered parent C and the purple-flowered parent B.

Since these progeny genotypes all have both alleles (P and p), they all have the heterozygous genotype. Since purple flower color is dominant and white flower color is recessive, and since 1/4 + 1/4 + 1/4 + 1/4 = 1 = 100%, there is a 0% chance of the progeny of true-breeding white-flowered and true-breeding purple-flowered parents to produce a white-flowered progeny. They will be 100% heterozygous and 100% with purple flowers since purple flower color is dominant to white flower color. Thus, all $F_1$ generation progenies have the same genotype (heterozygous) as well as the same phenotype (purple flowered). This explains why Mendel didn't find any white-flowered pea plants in his $F_1$ generation.

So as Mendel discovered, the allele that confers purple flower color masks the expression of the allele that confers white flower color. As we've seen, this kind of relationship between alleles in which one completely masks the expression of the other in the heterozygous condition is called *complete dominance.*

Mendel must have been amazed and excited when he found that white color in pea flowers reemerged in 25 percent of the *second generation*, or $F_2$ *progeny,* when $F_1$ generation plants were crossed. This clinched the argument that heritable factors can *maintain their integrity* from generation to generation. The genes for white flower color *had not disappeared* or melted into purple. They were indeed still in every cell of the $F_1$ generation and transmitted to the $F_2$ progeny. Here is the Punnett square that explains Mendel's data for his $F_1$ generation cross that produced $F_2$ progeny:

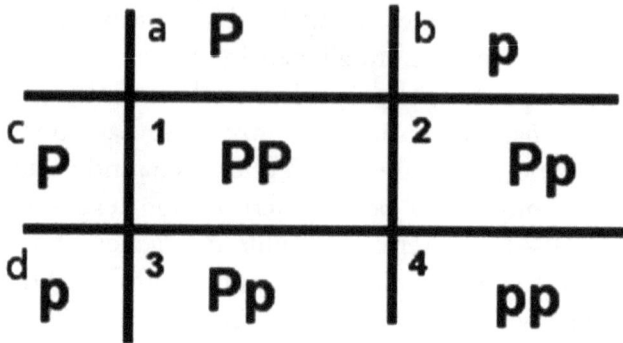

Figure 2.3. Punette square showing Mendel's F1 cross and all possible combinations of F1 alleles in the F2 progeny of that cross. These second generation progeny of the parental generation are called the F2 generation. F1 alleles are labeled with letters. F2 progeny genotypes (alleles from both F1 parents) are labeled with numbers.

The genotypic ratio, or ratio of kinds of genotypes in the $F_2$ progeny shown in boxes 1–4, are 1:2:1 (1 homozygous dominant {box 1} to 2 heterozygous {boxes 2, 3} to 1 homozygous recessive {box 4}). Because of the complete dominance relationship between these alleles, the phenotypic ratio is 3:1 (3 purple flowers {boxes 1–3} to 1 white flower {box 4}) or 75%–25%, matching the ratio that Mendel observed in his experiment described above. To repeat, the Punnett square, adapted from a statistical method used by Mendel, charts all possible allele combinations from the parents and, from these allele combinations, predicts the proportion of progeny that will possess each of the possible resulting genotypes.

We now know that this heritable factor, which maintains its integrity through generations, is the long-coveted mechanism of evolution that Darwin was looking for. In fact, Mendel had sent a copy of his scientific publication to Darwin; but Darwin, who wasn't too good at math, either never read it or didn't understand it (Henig 2000). Every living organism has heritable factors, or genes, in every cell. Thanks to updated microscope technology in the latter part of the nineteenth century and subsequent discoveries in the twentieth century, we now know that these genes, whose behavior was discovered by Mendel, occur in sequences along *chromosomes* (figure 2.4).

Genes

**Figure 2.4. Simplified diagram of a chromosome. Each shaded segment represents a heritable factor, or gene. In reality, a typical chromosome is composed of thousands of genes.**

The diploid number refers not only just to the number of each kind of gene in a living organism's cells but also to the number (two) of each kind of *chromosome* in a living organism's cells (figure 2.5). In other words, we inherit one of each kind of chromosome from each parent. Each kind of chromosome consists of a unique sequence of genes. These chromosome pairs—each chromosome with an array of the same kinds of genes as the other in the pair and each from a different parent—are called homologous chromosomes.

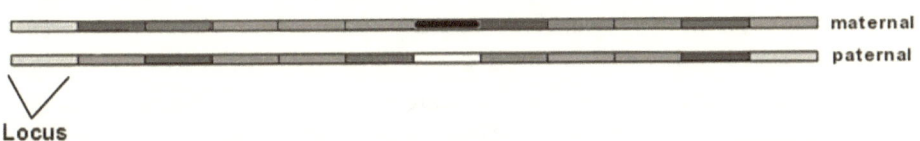

maternal

paternal

Locus

**Figure 2.5. A homologous pair of chromosomes, one from the maternal parent, the other from the paternal parent. A "locus" is a site on the homologous chromosome pair where a specific kind of gene occurs. Different shades at a locus indicates different alleles, or versions of the gene specific to that locus. When there is more than one allele at a locus, as depicted here, the locus is "heterozygous." If both genes at a locus were the same shade, this would symbolize that there is only one allele at that locus. The locus would then be "homozygous."**

# Where does all the diversity that natural selection works on come from?

*Mutations and alleles*

Like other intelligent social species of animals, we know members of our own kind better than we know others. We can recognize individual humans by their distinctive faces, sizes, shapes, and other features. Even though most of us aren't as good at this with other species of living things, almost all species have unique individual variation. Certainly, all sexually reproducing species do. This is fortunate, as this individual variation within a species is an important factor in a species's ability to persist and not go extinct if and when environmental conditions change. Take for example those rabbits. Luckily, they had some genes for the white color; otherwise, it would have been the end of the line for all of them when the ice age hit, even though those genes didn't serve them well before the ice age hit. From where does this variation in a species or population within a species arise?

At the turn of the twentieth century, just a few years after Gregor Mendel's work had been discovered, a geneticist named Thomas H. Morgan at Columbia University and his graduate student, A. H. Sturtevant, were rearing small fruit flies (*Drosophila melanogaster*) by the thousands. They irradiated their flies with x-rays to increase the rate of mutations. Some of the mutations were easily observable (e.g., white eyes instead of normal "wild type" red ones, wrinkled wings instead of "wild type" flat ones, and completely black bodies instead of "wild type" light brown ones) and didn't kill the fly. Morgan and Sturtevant found that these mutations were passed from generation to generation. They realized that they had altered the heritable factor in some of their flies. They had induced mutations in some genes. These genes had become altered into new versions, or alleles, and had increased the genetic (genotypic) diversity in their fly population, which resulted in the emergence in phenotypic variation, or observable diversity.

We now know that mutations are caused when the cell machinery that helps the chromosomes to reproduce inside a cell makes a mistake while a given gene on the chromosome is being replicated. X-rays can cause this, as well as a lot of other factors in the body and in the environment. Most of the time, these mutations are harmful—even lethal, if unchecked by the cell. But sometimes their effect is neutral or even helpful. For example, because of the way that the immunodeficiency virus (HIV) reproduces in human blood cells, its genes mutate frequently. While this causes quite a bit of viral death, some new mutations also allow the virus population to develop its own new strains of itself that survive and avoid the medicines we attempt to throw at HIV. This explains why cocktails of drugs are used and why researchers constantly work to develop new drugs to fight the AIDS virus. The AIDS virus evolves very quickly into new strains. Not good for HIV-positive people, very good for HIV viruses—all because of the ability of the HIV to mutate as frequently as it does.

So mutations that are beneficial and become new alleles are important in enhancing the genetic variation in populations of organisms. But how are these new mutations, or any and all alleles for that matter, new or old, mixed and shuffled into a population? Why is there so much diversity within a species? Why is DNA fingerprinting such a powerful kind of evidence in the modern courtroom, zeroing right in on a given individual?

Of course, part of the answer is random mating. Most human societies have taboos against marrying close relatives because the practice has long been found to result in unhealthy children. If a person marries and has children with someone he or she is not closely related to, chances are the person will be much more genetically different from that someone than if they were closely related. Randomized mating by unrelated individuals results in healthy genetic diversity in people and other organisms. Nonrandomized mating, as in inbreeding, results in an increased frequency of genetic diseases. What causes this increased frequency of genetic disease? Many harmful mutations are caused by recessive alleles. Enough of these harmful recessive alleles are masked by healthy dominant alleles in most situations so that the individual is reasonably healthy and free of genetic disease. But if close relatives have children, there will be an increased level of homozygous genotypes in the children and an increased likelihood of genetic disease, since many more harmful recessive alleles are not hidden by dominant alleles in the homozygous condition. A situation in which the population number is very low and close relatives are mating (so there's no opportunity for true random mating) is called *inbreeding depression*. The overall health of individual progeny becomes depressed, and the rate of infant mortality rises. In various organisms, this is sometimes a step before extinction. It is also a reason, scientists realize, that all new species, like our own (*Homo sapiens*), must be descended from a sufficiently large population with good genetic diversity in it. Of course, this fact does not conflict with an informed literal interpretation of the Adam and Eve account (see chapter 1), although it does conflict with an assumption that the words in that account are to be taken in today's context.

But random mating alone doesn't answer the question, why is there seemingly endless diversity in the human population or any sexually reproducing population? The full answer involves the process of gamete (sex cell) production, a process called *meiosis*.

*Meiosis*

All cells in an organism, whether the organism is a plant, animal, fungus, algae, or bacteria, are capable of reproducing themselves by dividing. Cell division has a number of valuable functions for a living organism, not least of which is the growth of the organism. In normal cell reproduction or mitosis, the new (daughter) cells that are the products of a cell's division are genetically identical with one another and with the mother cell. Every chromosome in a daughter cell has an identically corresponding chromosome in its sister cell. There are thousands or hundreds of thousands of genes on each chromosome in a cell, depending on the complexity of the organism.

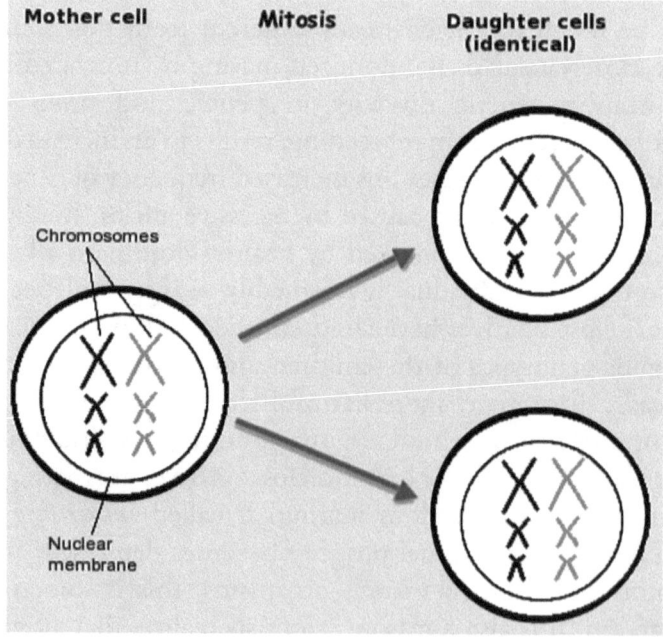

**Figure 2.6. Mitosis. The mother cell (left) divides into two genetically identical daughter cells. Mitosis occurs during normal cell division.**

JOSEPH FORTIER

Mitosis is conservative. Each cell division results in two genetically *identical* sister cells (figure 2.6.). The genetic information in the parent cell (the one that divides) is *precisely conserved* in each daughter cell. In contrast, meiosis is a process of cell division in which the four cells that are produced in this case are all quite *genetically different* from each other. Unlike simple mitosis, during meiosis the genetic information is shuffled and mixed. The products of cell division by meiosis, the eggs and sperm, both collectively called the "gametes," are all very different genetically from their mother cells, their grandmother cells, and from one another. What generates this genetic diversity? Actually, Mendel had some ideas that turned out to be important clues as to how some of this variation is generated by meiosis (the process of cell division resulting in gametes): his *law of segregation* and *law of independent assortment*.

The law of segregation is illustrated in the way the Punnett square is set up. A Punnett square is a tool used for calculating probabilities for any given combination of alleles (gene versions) donated by each parent for any given gene locus on a homologous chromosome pair (see Figs. 2.2, 2.3, and 2.5. In most sexual organisms, there are one or two possibilities for how many alleles there might be at that locus. As we saw, at homozygous loci, in which the same allele (gene version) is present on both paternal and maternal chromosomes for a given gene, there are two copies of one allele (as in the parental generation in the first Punnett square above). For heterozygous loci (such as those in figure 2.5), there is one copy of *each of two* alleles (as in the $F_1$ parents in Fig. 2.3 above and offspring 2 and 3). The law of segregation states that before gametes are formed, the two genes at a locus separate, or segregate, from each other. Thus, according to this law, each of these similar genes—one inherited from the father and the other from the mother—in all likelihood will not wind up in the same egg or sperm cell. We'll look at this closer below. The law of independent assortment is a little more involved. It states that for any two or more loci (each locus is a homologous pair of genes as in figure 2.5), if a gene from one locus, paternally or maternally inherited, winds up in a given gamete, this has no influence on which gene (paternally or maternally inherited) of any other locus will wind up in that same gamete.

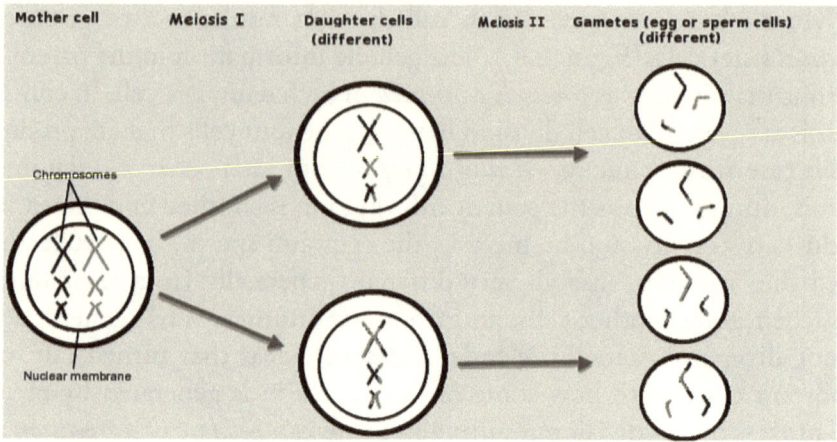

**Figure 2.7. Meiosis. Two cell divisions of the mother cell give rise to the gametes. Before and during the first division, or meiosis I, the genetic material is shuffled twice, and the deck cut. During Meiosis II, the deck is cut again, and the hand is dealt. The resulting genetic diversity is enormous, even among the gametes of an individual organism. Meiosis occurs during the process of egg and sperm (gamete) formation. Black represents paternally inherited genetic material. Gray represents maternally inherited genetic material.**

In order to make these laws clearer and to see what they have to do with generating genetic diversity and variation, we need to look at the process of *meiosis*: the process of cell division that leads to the formation of *gametes*, or egg and sperm cells, in plants, animals, many fungi, and many microscopic single-celled creatures, including algae (figure 2.7).

We will focus in on what happens to chromosomes and genes during sex cell formation. We'll do this in order to see more clearly how a diverse pool of gene versions (alleles) to draw from, in order for natural selection to operate, is generated and maintained in populations.

Before meiosis starts, the chromosomes *replicate themselves* in the cell that is about to undergo meiosis. Large complex molecules reproduce themselves. Amazing, and also true. In chapter 3, we'll see how chromosomes actually do this.

As the cell machinery organizes for meiosis, the chromosomes—which up to now have been thin molecular strands in the cell, invisible under a microscope—start to become visible under the scope as they kink up, appearing shorter and thicker. Each shortened, thickened chromosome looks similar to one of the six chromosomes in figure 2.7, left side of figure. At this stage, each chromosome is composed of two copies of the original chromosome before replication. The process begins with a mother cell that gives rise to four gametes, which are egg cells or sperm cells, depending on the sex of the individual in which meiosis is taking place (figure 2.7).

There are two major cell divisions that occur during meiosis. During the first division, most of the genetic shuffling takes places among the chromosomes. During the second division, another shuffling event takes place, and the gametes are formed.

The first two shufflings of the genetic material on the chromosomes that happen during meiosis occur in the mother cell before it divides. For each chromosome type in a cell, there are two homologous chromosomes: one that was inherited from the father (paternal chromosome) and one from the mother (maternal chromosome) (figure 2.7, left side). For each chromosome type, all chromosomes carry genes that code for the same instructions used to build proteins used by the organism. However, as we've seen, there are variations or versions of each of these genes. Each version, or allele, codes for a variation in the structure of the protein that that particular gene codes for, which in turn causes a slight difference in the expression of the protein. Thus, an allele of the gene that codes for ear shape in humans codes for ears with earlobes on the bottoms of the ears. Another allele, or variation of that same gene, codes for ears without earlobes (figure 2.8). In humans, there are twenty-three types of chromosomes. For each type, one is paternal, and the other maternal. Humans normally have forty-six chromosomes.

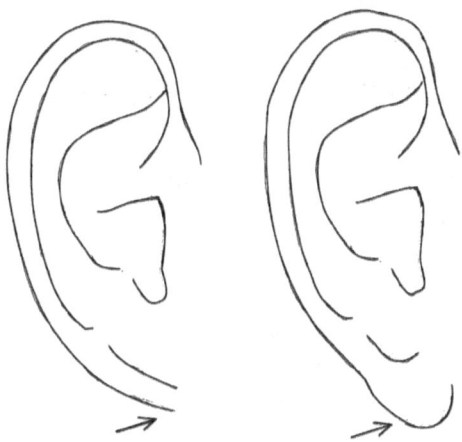

**Figure 2.8. Two alleles that occur at the same locus on a chromosome determine whether the human earlobe is attached (left) or lobed (right).**

The first major shuffling between the chromosomes of a homologous pair occurs when the paternal and maternal chromosomes pair up and exchange genetic material (figure 2.9). The result is that when they separate from each other—where before, each chromosome had been completely composed of *either* maternal *or* paternal genes—now each is composed of *a patchwork of* maternal and paternal genes.

This genetic shuffling is a major source of genetic variation. Since some combinations of alleles lend greater adaptive ability in a given set of environmental conditions than other combinations, natural selection can work on this variation. Allele combinations that confer greatest ability to survive and reproduce under a given set of environmental conditions will tend to be retained in a population of organisms over generations.

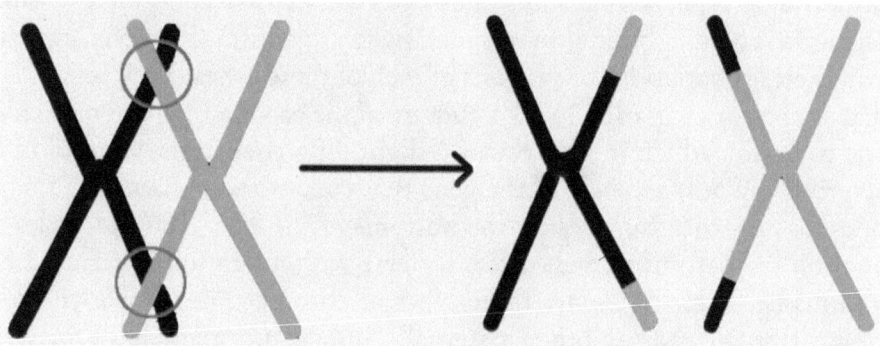

Figure 2.9. Crossing over. On the left are two homologous chromosomes of the same type and same shape. The black-colored chromosome is paternally inherited, and the gray-colored chromosome is maternally inherited. Circles show where the shuffling of the genetic material (crossing over) occurs between the chromosomes. On the right are the two chromosomes after this first shuffling event of meiosis I.

Another important source of genetic shuffling occurs during meiosis I. Mendel described its effects on pea color as the law of independent assortment. After the homologous chromosomes have paired, joined, and exchanged genetic material, they separate again but remain close to each other. These pairs then line up on an imaginary line in the middle of the cell, which is called the metaphase plate (figure 2.10). The rub here is, on which side of the imaginary line for a given homologous chromosome pair does the maternally derived chromosome come to lie, and on which side does the paternally derived chromosome lie? For any given homologous chromosome pair, whether a maternally derived chromosome or a paternally derived one is on a given side of the metaphase plate is a *complete and total flip of the coin.* If there are two people, each with a coin, and they both flip their coins simultaneously, the chance of either coin coming up with heads or tails is completely *uninfluenced by the other coin.* The coins assort themselves independently of each other. In the same way, using Mendel's parlance, how a given pair of heritable factors (heads or tails = maternal or paternal) orient themselves with respect to the metaphase plate is *independent* of how any and all *other* such homologous chromosome pairs assort *them*selves. Of course, metaphase plates and homologous chromosome pairs were unknown to Mendel. They weren't discovered until the 1870s and 1880s, when vast improvements in the optics used for microscopes allowed biologists to observe the events of meiosis that we are now discussing. The amazing thing is that Mendel's laws of heredity, derived from his observation of ratios of phenotypes in pea plant progeny with the help of combination theory, *were corroborated* by what later biologists *actually saw happening* in the behavior of these heritable factors.

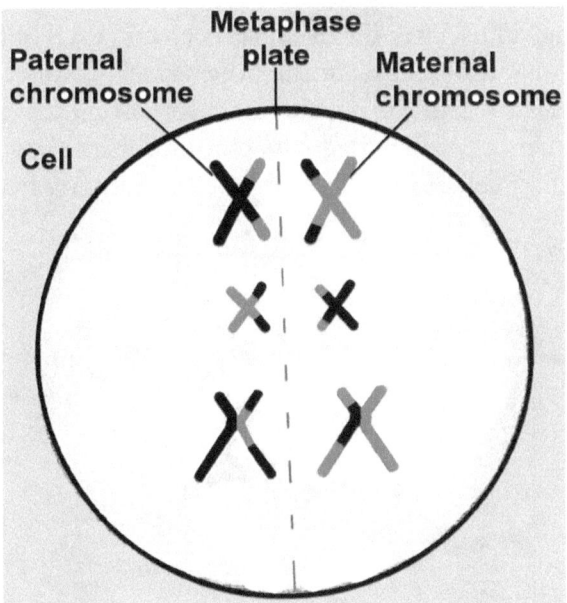

**Figure 2.10. Independent assortment of chromosomes during meiosis I after homologous chromosomes exchange genetic material during crossing over. Homologous chromosome pairs have lined up along the imaginary metaphase plate.**

The next event in meiosis I was anticipated by Mendel's law of segregation. According to this law, heritable factors separate (segregate) during gamete formation. This law nicely describes what happens next after independent assortment of the respective homologous chromosome pairs, each composed of a maternal and a paternal chromosome, on the imaginary metaphase plate. After this, the cell elongates, and each chromosome in a homologous pair separates from the other. They move to opposite ends of the elongated cell (figure 2.11). The genetic deck has been shuffled during genetic exchange and independent assortment. In this stage, the deck is cut. Now each end of the elongated cell (ready to divide) has some combination of one of each chromosome (maternal or paternal) from each of the homologous pairs. Note that each paternal and maternal chromosome is actually a patchwork of sections of paternally and maternally derived genetic material at this stage (figure 2.11), due to the genetic exchanges between chromosomes of each homologous pair illustrated in figure 2.9. The elongated cell then splits in two and divides into two daughter cells (figure 2.7, middle). Each of these daughter cells is unique, genetically different from the other daughter cell and from the mother cell due to genetic shuffling. Thus, after the deck has been cut, each of the two "stacks of cards" is unique, different from the other "stack." Notice also that each daughter cell has only half the number of chromosomes as the mother cell. Unlike the mother cell, which had *both* chromosomes of each homologous pair, each daughter cell *only has one* of the chromosomes from each pair.

**Figure 2.11. After independent assortment of chromosomes, the cell elongates, and homologous chromosomes draw apart from each other and move toward opposite ends of cell.**

JOSEPH FORTIER

During meiosis II, the formation of egg or sperm cells, or gametes, is finalized. The main event during meiosis II occurs when each daughter cell elongates. As we saw above, at the beginning of meiosis I, each chromosome was composed of two copies of the original chromosome before that original chromosome replicated itself. For each of those chromosomes, during the first shuffling event of meiosis I, one replicated half of a chromosome in a homologous pair–exchanged genetic material with a replicated half of the other chromosome in the pair (figure 2.9). In meiosis II, the two replicated halves of each chromosome break apart from each other, and each half moves to the opposite pole of the elongated cell with respect to the other half (figure 2.12, top). The genetic deck has now been reshuffled. Each of these two elongated cells then divides in two, yielding four genetically unique gametes (figure 2.7; figure 2.12, bottom).

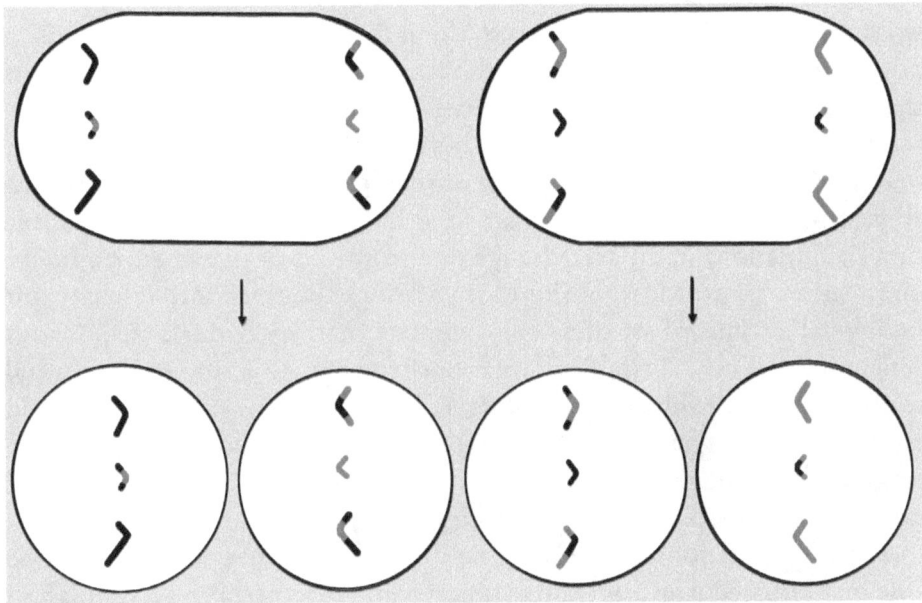

**Figure 2.12. Meiosis II.** At the top of the diagram, the two daughter cells (products of meiosis I) are elongating and the chromosomes in each have broken apart into their two halves. Each of these halves, which has migrated to an opposite end of the cell from its other half, is now considered an independent chromosome. Each half carries slightly different genetic information from the other. As can be seen, one of these halves from each formerly unbroken chromosome has both maternal and paternal genetic material, thanks to the first shuffling event of meiosis I (crossing over; Figure 2.9).

Thus, by the process of meiosis, two cell divisions occur, beginning with a mother cell, in which the heritable factors, or genes, are shuffled and reshuffled. The products of meiosis for a given mother cell are four genetically distinct gametes, or reproductive cells. Given the immense number of possible outcomes from the genetic-shuffling events of meiosis, the probability that any two of the thousands of gamete mother cells produced by an organism over its lifetime will produce genetically identical gametes is vanishingly minute. The capacity for meiosis to produce genetic diversity is astounding.

## *Conclusions*

We have seen that evolution is a process that occurs over generations. During this process, natural selection caused by environmental conditions working on heritable traits affects the survival of individuals within a population differently. Individuals that fortuitously have inherited traits that help them to survive best under the given environmental regime (rabbits with white coats in snowy environments) are most likely to live to reproduce; those without enough of such traits (rabbits that stay brown all year in snowy environments) are least likely to live to reproduce. Over time, populations in an area change due to the effect of the environment on inherited traits. Much of this change brings about traits that better suit individual organisms to their environment than individuals in previous generations. Thus, helpful adaptations emerge as a result of natural selection. As we will see in chapter 4, in populations of the same kind or *species* of an organism in which, for example, two populations remain separated for enough generations, they may become new kinds or species of organism. After separation of two such populations over many generations, these two populations may become so different that they will no longer be able to interbreed if and when they again come together. The accumulation of genetic differences between two populations isolated from each other in response to the respective environmental pressures playing on heritable traits in each population is responsible for such speciation events.

We've also seen that the mechanism for evolution is the heritable factor, or gene, first described by Gregor Mendel in 1866 from its pattern of expression in pea plants over generations. Genes are arrayed in sequence on long chromosomes, which are found in every cell in a living creature's body. For sexual organisms, chromosomes occur in pairs: one inherited from the father and the other from the mother. Thus in humans, there are forty-six chromosomes composed of twenty-three pairs of similar or homologous chromosomes: one paternally derived, the other maternally derived. Each of the chromosomes of a homologous pair has a sequence of genes that matches the sequence on the other in the pair. The site of a specific gene on a chromosome is called a locus. At a given locus of a pair of homologous chromosomes (figure 2.5), there may be two versions of that gene: a different version, or allele, on one chromosome from its homologue. These different alleles allow for the genetic diversity in populations that can be acted on by natural selection. Under a given set of environmental conditions, some alleles become more common and others less common due to differential survival and retention in the survivors of the alleles that confer superior adaptive traits, such as alleles that are expressed as white coat color in rabbits that live in snowy places.

Genes are passed on from parents to offspring in egg and sperm cells, or gametes. These sex cells are produced by meiosis, a process that ensures the genetic diversity of the offspring. During this process, genes and chromosomes are shuffled during divisions of the cell lines that give rise to the gametes. Heritable genetic diversity allows for expression of a diversity of traits in the offspring. Within each generation, no two individuals are completely alike. This variation among individuals is always tested by environmental conditions. Some of these conditions change quickly over time, others slowly. This individual variation—born of the genetic diversity generated by mutations, random mating, and meiosis—ensures that some members of a population will survive to reproduce and that the population will live on for generations. It also explains the way the heritable mechanism, or gene with its diversity of alleles, works in response to environmental factors. By adapting to these environmental factors, over generations a species of living organism changes. In other words, descent with modification or evolution occurs, caused by adaptation to the environment by natural selection working on heritable traits.

# Summary

☐ Science uses rationality and observation of empirical data to answer questions. Preliminary conclusions are *hypotheses*. When repeated observation and experimentation consistently confirm a hypothesis, it becomes a *theory*.

☐ Hypotheses and theories may be accepted or rejected based on new insights from empirical observation of the data. If accepted, they are sometimes later modified or refined based on insights gleaned from subsequent data.

☐ Through repeated observation of similar patterns in many types of organisms in many different places, including Galapagos mockingbirds, Darwin reasoned that evolution, or *descent with modification*, is caused under natural conditions not influenced by human manipulation but by *natural selection* working on heritable traits in populations of organisms, culling those whose heritable traits are less *adaptive* to environmental conditions. His reasoning based on the data has stood the test of constant testing by skeptical scientists over time.

☐ The missing smoking gun or mechanism for descent with modification was discovered by Gregor Mendel but was not recognized by others until 1900. His *heritable factor* was later identified as the *gene*.

☐ Sequences of genes occur on *chromosomes*. Each kind of chromosome occurs in all cells of an organism. For each kind of chromosome in sexual organisms, there are two varieties—one inherited from the father, the other from the mother. These chromosomal pairs are *homologous chromosomes*. For a given *locus*, or site on a homologous chromosome pair, there may be one or two versions, or *alleles*, of a gene.

☐ In the formation of the *gametes*, during the process of *meiosis*, a great deal of shuffling and mixing of genetic diversity takes place between chromosomes.

☐ Each grandmother cell of the gametes gives rise to four gametes (egg cells or sperm cells).

☐ Mendel's *laws of segregation and independent assortment*, which he realized were together responsible for mixing and shuffling alleles in a random, chance fashion, have been independently verified by later scientific observers working out the events of gamete formation (meiosis) in cells using high quality microscopes and lab equipment that were not available in Mendel's time.

☐ The genetic diversity generated by *mutations, random mating,* and *meiosis* results in individual variation in populations of organisms. Natural selection acts on this variation. Natural selection is the genetic pressure put on a population of organisms by environmental factors. A healthy population with plenty of genetic diversity (lots of alleles, or versions of each gene) is capable of responding to this natural selective pressure. Individuals best suited to survive and reproduce in an existing environmental regime by their inherited traits, coded for by their genes, tend to pass on their genes to future generations more frequently than those less suited. They are better adapted to their environment. Over generations and time, a genetically healthy population of organisms changes genetically and changes again, constantly adjusting and readjusting to its changing environmental regime. Natural selection is the engine of genetic change over time most responsible for conferring the fitness in the population that helps it to persist. Thus, natural selection works to build more frequent occurrence of genes that confer adaptations in the population that in turn confer better fitness for the environment in which the population's members find themselves in any segment of time.

# Glossary

**adaptation**. The result of increased fitness in a population of organisms to a given environmental regime in a given period of time due to natural selection. Also, a specific heritable trait of organisms in a population that confers survival value for a given environment or way of life in that environment.

**allele**. A version of a gene, often conferring a different quality of expression in some way than other versions of that same gene.

**chromosome**. A long macromolecule of DNA with its associated proteins and other molecules that assist in the functioning of the DNA. A chromosome is composed of a sequence of genes, each of which in turn is composed of a segment of DNA and associated proteins and molecules.

**complete dominance**. A genetic system consisting of one gene with two or more versions (alleles) in which at least one allele entirely masks the expression of one or more other alleles of that gene.

**descent with modification**. The process of evolution in which a population of organisms changes over generations and new species emerge from ancestral species due to natural or artificial selection and other forces that interact with inherited characteristics of individuals in the population.

**diploid**. The number of chromosomes in each cell of individuals of those kinds of organisms in which both maternally and paternally inherited chromosomes are present. For each kind of chromosome and thus each kind of gene on a given chromosome, there is a pair: one member of the pair paternally inherited and the other maternally inherited.

**dominant gene** (or **dominant allele**). A version of a gene that, when paired with another version of that same gene after a fertilization event, masks the expression of the other gene and is itself fully expressed.

**empiricism**. A theory of knowledge that knowledge is dependent on sense experience.

JOSEPH FORTIER

**F$_2$ progeny.** In an experimental fertilization trial, the offspring of fertilization events between the offspring of a parental cross; the second generation after a parental cross.

**gamete.** A reproductive cell; an egg cell or a sperm cell.

**gene.** A functional unit of a chromosome which has a discreet function, affecting or determining a discreet trait of the organism. Genes occur in sequences along a chromosome.

**genotype.** The sum total of gene versions, or alleles, present in a given individual organism. Alternatively, for a given gene, the alleles present in the organism for that particular gene.

**heritable factor.** The substance in living organisms that expresses itself as a given trait or characteristic. It was discovered that it is synonymous with the gene.

**heritable traits.** Quantitative or qualitative characteristics of organisms for which the instructions are coded in the genetic material of each organism, expressed as these traits, and capable of being passed on to offspring.

**heterozygous.** The condition in diploid organisms in which two alleles for a given gene are present in the organism's genome.

**homologous chromosomes.** Chromosomes that occur in pairs or other even numbers in each cell of a diploid organism, one or more being inherited from the organism's mother and an equal number from the organism's father. Homologous chromosomes are all composed of sequences of the same genes, although each chromosome in a homologous grouping may be composed of different alleles of a given gene from other chromosomes in that homologous grouping. For most sexual organisms, there are only two homologous chromosomes, one inherited from the organism's female parent and the other from the organism's male parent.

**homozygous.** The condition in diploid organisms in which only one allele is present for a given gene in the organism's genome; in other words, both maternally inherited and paternally inherited genes are the same version.

**hypothesis**. A provisional assertion that can be tested, with the possibility of being falsified, by scientific investigation, including experimentation.

**inbreeding depression**. A condition that may arise in populations of sexual organisms in which the population number is so small that close relatives are breeding. Because of the low population number, the population's genetic diversity (number of alleles per gene) is also low. The overall effect is an increase in homozygous genotypes of offspring over generations, which allows for an increase in frequency and number of harmful recessive genotypes to be expressed, and decreases (or depresses) the overall adaptive fitness of the population, often driving populations to extinction.

**law of independent assortment**. A law of Mendelian heredity that states that for any two or more homologous chromosome pairs, if a chromosome of one homologous pair, paternally or maternally inherited, winds up in a given gamete, this has no influence on which chromosome (paternally or maternally inherited) of any other chromosome pair will wind up in that same gamete.

**law of segregation**. A law of Mendelian heredity that states that before gametes are formed, the two genes at a locus on homologous chromosomes separate, or segregate, from each other.

**locus**. A place or site on a chromosome where a given gene is located.

**meiosis**. The process of gamete formation in organisms in which the genetic material is shuffled and mixed over two cell divisions to yield four genetically unique gametes.

**metaphase plate**. An imaginary plane in the center of a dividing cell, during either mitosis or meiosis, on which pairs of homologous chromosomes align themselves, such that one chromosome of the pair is on each side of the metaphase plate.

**mitosis**. The process of normal cell division in which a mother cell divides to form two daughter cells. The daughter cells are *genetically* identical to one another and to the mother cell.

**mutation**. An abnormality in a gene caused by a mistake that was made when it was replicated from an ancestral gene. Most mutations result in traits that have either a neutral or a harmful effect on the organism; however, some mutations may be or become beneficial.

**natural selection.** The process by which individuals in a population that are less genetically fit for the natural environmental regime during a given time period are culled from the population by mortality before they reproduce, and individuals that are more genetically fit for that environmental regime during that given period survive to reproduce.

**parental cross.** In an experimental fertilization event, fertilization between one or more female individuals for which the genotype for a certain trait is known to be homozygous, and one or more male individuals for which the genotype for that same trait is also known to be homozygous.

**phenotype.** The expression of a gene or group of genes, usually observable as a discrete trait.

**random mating.** In populations of sexually reproducing organisms, the theoretical freedom of an organism to reproduce with any organism of the opposite sex in that population.

**realism.** A philosophical position that holds that reality is independent of our conceptual schemes and that during any given time period, how we see things is only an estimation of reality, to which we approximate more closely by the process of learning from observation.

**recessive gene** (or **recessive allele**). A version of a gene that, when paired with another version of that same gene after a fertilization event, is inhibited from being expressed by the other gene version.

**species.** A genetically unique kind of organism. In sexual organisms, a species is best defined as a group of organisms in which matings or fertilization events between males and females give rise to healthy offspring capable of themselves giving rise to healthy offspring and, furthermore, regularly mate in their natural environmental setting.

**theory.** A hypothesis that has come to be generally accepted by the scientific community due to repeated testing and consistent affirmation.

# References

Darwin C. 1859. *Origin of Species*. London: John Murray

Gray D. 2007. Ukaliq: the Arctic Hare. Edited by K. Quinn. Ottawa: Canadian Museum of Nature Last modified February 16, 2011. http://nature.ca/ukaliq/021des/100_nmc02_e.cfm

Henig R. M. 2000. *The Monk in the Garden*. New York: Houghton Mifflin Co.

Kripilani J. B. 1970. "Gandhi: His Life and Thought." Publications Division, Ministry of Publication and Broadcasting, Government of India

Markie P. 2008. "Rationalism vs. Empiricism." In *The Stanford Encyclopedia of Philosophy (Fall 2008 edition)*. Last modified August 2008. http://plato.stanford.edu/archives/fall2008/entries/rationalism-empiricism/

Miller A. 2008. "Realism." In *The Stanford Enyclopedia of Philosophy (Fall 2008 Edition)*. Last modified 2008. http://plato.stanford.edu/archives/fall2008/entries/realism/

# CHAPTER 3

# A Closer Look at the
# Heritable Factor and How It Functions

WHAT IS IT about the gene that makes it the heritable factor, the mechanism of inheritance and evolution in all living things? Is there a basic ingredient that confers this seemingly mysterious property on the gene? How does the gene store and reproduce information that is passed to subsequent generations? In short, how does the mechanism *work*? In this chapter, we'll take a look at the history of discovery of the answers to these questions—questions that tantalized geneticists, evolutionary biologists, and biochemists for the first five decades of the twentieth century. In the process, we will answer these questions.

## Friedrich Miescher and the discovery of DNA

In 1868, a Swiss biologist named Friedrich Miescher, a contemporary of Mendel's, found that the nuclei of pus cells (white blood cells) have a phosphorus-containing substance, which he named nuclein. Miescher found that this substance has an acidic portion and a protein portion. Later, he found the same substance in the sperm cell heads of a salmon (Biotechnology Industry Organization 1990). The acidic portion of nuclein became known as deoxyribonucleic acid, or DNA for short, and was found to be rich in phosphorus. The protein portion, also part of the chromosome, is now known to assist the DNA functioning in various ways (Stent 1980, xiii–xiv).

## Early twentieth-century work on DNA chemical composition

By the early 1900s, biochemists had found that DNA occurs in cells of all plant and animal species tested for its presence. They also found that DNA is a gigantic molecule, or *macromolecule*, composed of three different kinds of smaller molecules. One group of these smaller molecules are *nitrogenous bases*. There are four kinds of nitrogenous bases. A second kind of smaller molecule that composes DNA is the five-carbon sugar *deoxyribose*.

Finally, the third small molecular component of DNA is phosphoric acid, or *phosphate* (this is where the phosphorus that Miescher detected in DNA is found). The basic building block of DNA was found to be composed of (a) one molecule of any of the four kinds of nitrogenous bases, (b) one molecule of deoxyribose sugar, and (c) one molecule of phosphate.

There are some common characteristics of organic macromolecules produced in living organisms that will make our discussion of DNA a bit more understandable. First off, DNA is a *polymer*. A polymer is any kind of a macromolecule composed of a chain of links in which each link is a molecular subunit, or *monomer* (figure 3.1). Thus, the polymer DNA is composed of a chain of *nucleotide* monomers (figure 3.2) (Stent, xiii–xiv). Now since there are four kinds of nitrogenous bases, there are four possible (and as it turns out, actual) kinds of nucleotides, depending on which kind of base is in any given nucleotide (figure 2). By the late 1920s, it was known that DNA occurs almost exclusively in the chromosomes.

**A polymer macromolecule**

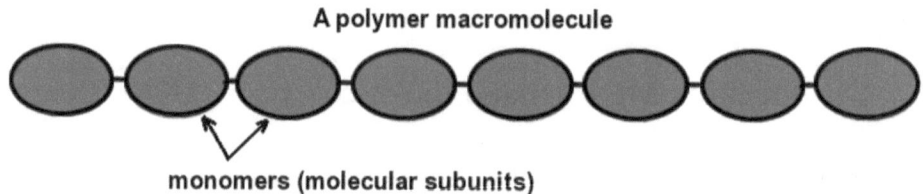

**monomers (molecular subunits)**

Figure 3.1. A generalized polymer composed of a chain of smaller molecular subunits, or monomers.

## Frederick Griffith: is DNA the heritable factor component of chromosomes?

In 1928, Frederick Griffith, a British biologist working with the British Department of Health, published the results of his experiment that provided evidence that DNA may be the heritable factor. Despite the frugality of the Department of Health in equipping labs at the time, Griffith was incredibly imaginative and resourceful. In his experiment, Griffith used two strains of the bacterium *Streptococcus pneumoniae*, the bacterial species that causes pneumonia. One strain was smooth (S-strain). The outer *polysaccharide* coating of each bacterium looked smooth when viewed through a microscope. A polysaccharide is a macromolecule in which the monomer is some type of sugar. This was the virulent strain—it caused death by pneumonia in lab mice when it was injected into them (figure 3.3A). The other strain, the rough strain (R-strain), had a polysaccharide coating that appeared rough when viewed through a microscope. When lab mice were injected with this strain (figure 3.3B), they neither showed symptoms of illness nor died.

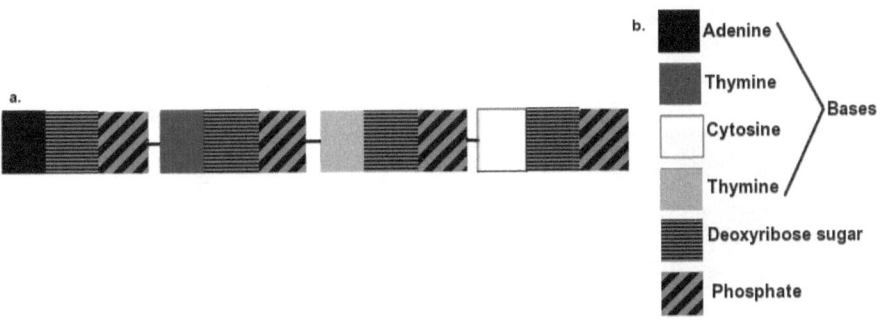

Figure 3.2. Early conception of DNA. A. section of DNA polymer. Each subunit, or monomer, is in turn composed of (1) one of four kinds of base molecule, (2) a deoxyribose sugar molecule, and (3) a phosphate molecule. DNA monomers are called nucleotides. B. legend of shading codes for nucleotide subunits.

Griffith, being a curious individual and not one for rash assumptions, next designed an experiment to test whether it was the coating in the smooth bacterial strain that caused disease *or something else*. He heated some of the S-strain bacteria, killing them, and injected the dead bacteria into mice (figure 3.3C). The mice lived. Since Griffith knew that heat does not *denature*—or change the chemical nature of polysaccharide—but *does denature protein*, he concluded that the outer coating *was not* the cause of disease.

This left something else as the suspect.

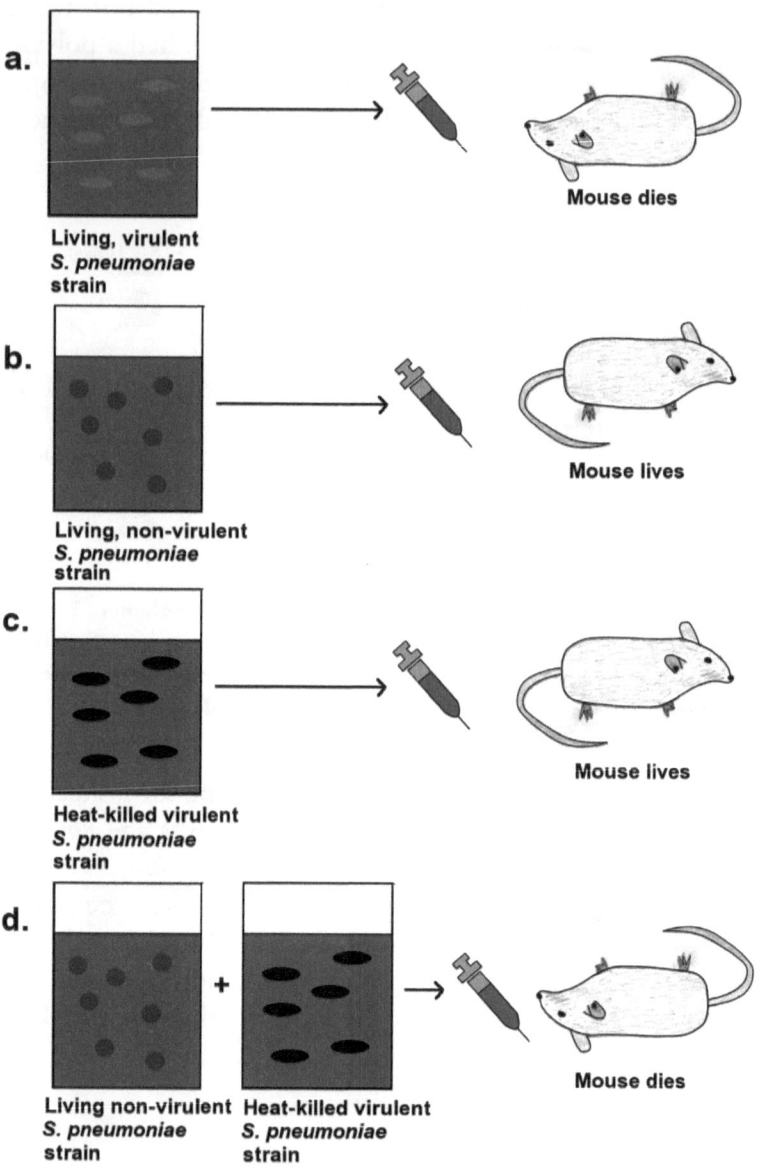

**Figure 3.3. Griffith's experiment using pneumonia-causing bacteria and lab mice. His experiment provided evidence that rather than protein, DNA is the heritable factor. A. When experimental mice were injected with live virulent *S. pneumoniae* bacteria, they died. B. When mice were injected with live non-virulent *S. pneumoniae* bacteria, they lived. C. When the mice were injected with heat-killed virulent *S. pneumoniae* bacteria, they lived. D. When the mice were injected with both live non-virulent *S. pneumoniae* bacteria and heat-killed virulent *S. pneumoniae* bacteria, they died.**

JOSEPH FORTIER

Finally Griffith injected both living non-virulent bacteria and heat-killed virulent bacteria into a fourth group of mice (figure 3.3D). They died. From these live mice, he isolated *live S-strain* bacteria. Somehow, the live R-strain bacteria must have absorbed the heritable factor from the dead S-strain corpses. So the heritable factor couldn't have been protein since the S-strain bacterial protein was denatured by the heat. It was already known that bacteria of different strains can undergo transformation. *Transformation* occurs when bacteria of a given strain absorb the genetic material from bacteria of another strain, and that absorbed genetic material expresses itself, transforming the recipient strain into the donor strain (Bookrags 1999; Carter 1996; Medawar 1980).

The only suspect left was DNA. Yet the scientific world remained skeptical. Many just could not believe that the heritable factor was not protein. DNA, from what was known of it, seemed unlikely (Carter 1996).

In 1952, Hershey and Chase clinched the identity of the heritable factor.

## The Hershey-Chase Experiment

Alfred Hershey and Martha Chase, both Americans, conducted an experiment at the Carnegie Institution of Washington at Cold Spring Harbor, New York. The experiment involved bacteria; *bacteriophages* (*phages* for short) (figure 3.4), which are *viruses* that attack bacteria; radioactive sulfur and radioactive phosphorus; and an ordinary kitchen blender. Recently, the structure, composition, and behavior of viral particles had been discovered using electron microscopy. Max Knott and Ernst Ruska, both Germans, invented the electron microscope in 1930. Soon after, it became known that viruses consist of a protein outer coat and either DNA or its close chemical relative, RNA, inside the coat. Viruses differ from living cells in that they have no protein machinery associated with genetic material. It was also known that viruses reproduce themselves from the information of their own genetic material that is injected into the cell of the host (figure 3.5). They use the host cell's protein hardware associated with the host's genetic material for their own reproduction. In a way, they are tiny pirates, invading the ship of the host cell and taking over its protein equipment for their own disreputable purposes. Eventually, many new phages burst out of the cell, killing it, to attack other host cells. The process is repeated as other host cells are invaded (figure 3.5).

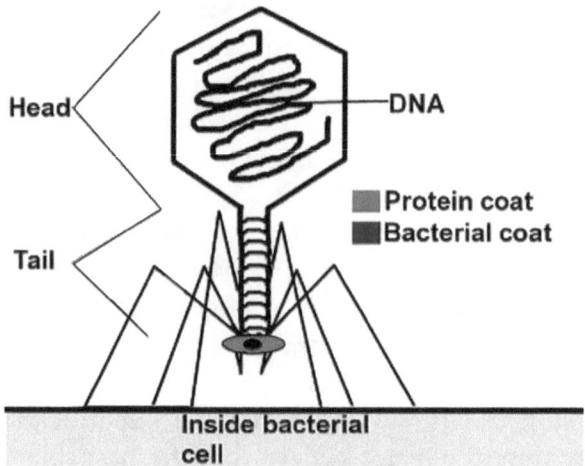

Figure 3.4. A bacteriophage, or bacteria-attacking virus, on a bacterial cell surface. The DNA will be injected into the bacterial cell, where it will commandeer the bacterial cell machinery to assist in bacteriophage replication.

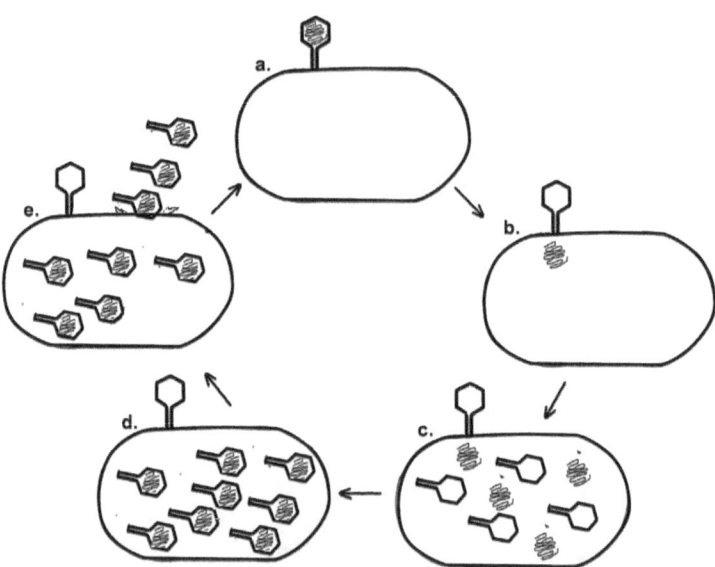

Figure 3.5. Life cycle of a bacteriophage. A. Phage locates the bacterial cell. B. Phage injects its DNA into the bacteria. C. Phage DNA replicates and commandeers cell to replicate new phage protein coats using host proteins. D. Phage particles assemble. E. Phage particles burst bacterial cell membrane, disperse to other bacteria, and cycle starts over.

JOSEPH FORTIER

The *T2 phage* was known at the time of the Hershey-Chase experiment. This type of virus was known to attack the bacterium *Escherichia coli* (*E. coli* for short) by injecting its genetic material into the bacterium. *E. coli* is a common intestinal bacterium in humans. Furthermore, it was known that (1) DNA contains phosphorus but no sulfur and (2) protein contains sulfur but no phosphorus. Hershey and Chase ran a parallel experiment (figure 3.6), one part involving sulfur and the other part phosphorus, in order to answer the question, is the genetic material injected into the bacteriophage DNA or protein?

The researchers allowed a colony of T2 bacteriophage to grow on a culture of *E. coli* bacteria that was being reared on food laced with radioactive sulfur but no phosphorus (figure 3.6, culture 1) and another colony of T2 phage to grow on an *E. coli* culture reared on food laced with radioactive phosphorus but no sulfur (figure 3.6, culture 2). They allowed enough time to pass for the viruses in each *E. coli* colony to go through a few reproductive cycles so that they could be sure that the viruses they were to collect would have absorbed one of the radioactive substances into their own genetic material: phosphorus, if the genetic material is DNA, or sulfur, if the genetic material is protein.

A T2 phage sample was collected from each bacterial culture. Each of those samples was placed in a fresh *E. coli* culture with no radioactivity and grown for only ten minutes—enough time for the phages to inject genetic material into the bacteria, but not enough time to reproduce new phages. The two batches of phage-infected bacterial cultures were each put in a blender and agitated to knock off viruses still attached to bacterial cells (figure 3.6). Each batch was then placed in a test tube and the test tube placed in a centrifuge. The centrifuge was spun at high speed to force the heavy parts of the batches to the bottoms of their respective tubes by centrifugal force. When the centrifuge was turned off, there was a thick *pellet* in each tube, composed of the heavy bacterial remains, and liquid *supernatant*, composed of liquid solution and the viral protein coats (figure 3.6B, C).

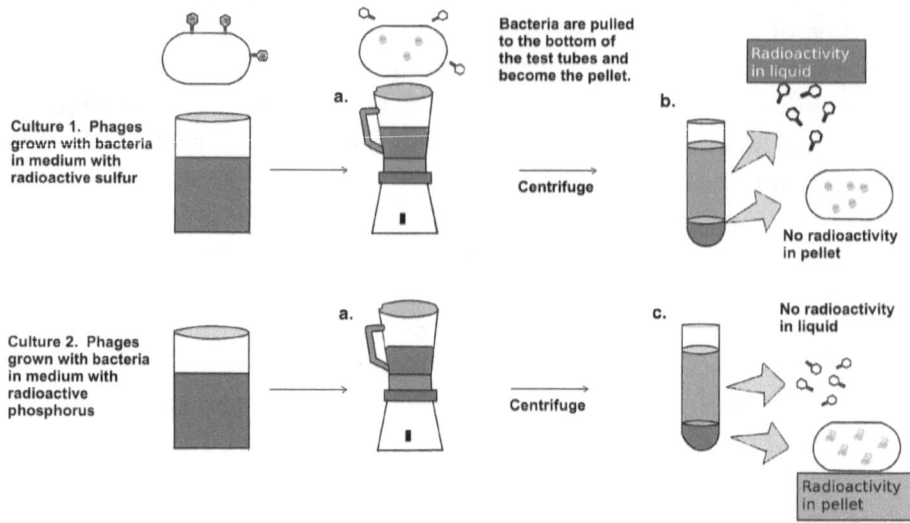

**Figure 3.6. The Hershey-Chase experiment.** Two parts of the experiment were run simultaneously. In each part, a culture of *E. coli* bacteria was infected with a culture of T-2 bacteriophage virus that had been previously exposed to a radioactive substance. In culture 1, the radioactive substance was sulfur. In culture 2, it was phosphorus. In each culture, viral DNA was injected into the bacteria by the viruses. In both cultures, after infection, the medium was spun at high velocity to separate bacteria from viral protein coats. Viral coating remained suspended in solution. Bacterial remains, including any injected viral DNA, was compressed into a pellet. In part 1, radioactive sulfur was present in the solution. In part 2, radioactive phosphorus was present in the pellet. Protein absorbs sulfur. DNA absorbs phosphorus. Part 1 demonstrated that protein could not be the heritable factor, since all radioactive sulfur was found in the liquid and not in the pellet. Part 2 demonstrated that the heritable factor is indeed DNA since radioactive phosphorus was found only in the pellet and not in the liquid.

Hershey and Chase found that in the test tube with viruses reared with radioactive sulfur (figure 3.6B), the liquid supernatant where the viral coats were was radioactive *and not* the pellet composed of bacterial remains. Thus, *no viral protein* had been injected into the bacteria. This amounted to strong evidence that protein is *not* the heritable factor. Conversely, in the test tube with viruses reared with radioactive phosphorus (figure 3.6C), the radioactivity issued from the pellet composed of dead bacteria and *viral DNA*. This, in turn, was strong evidence that DNA *is* the heritable factor. Mendel had found the smoking gun; Hershey and Chase found the bullet.

Why was this experiment so exciting to scientists? Finally, solid evidence was offered that corroborated previous studies suggesting that DNA is the heritable factor. The skeptics were convinced. The only radioactively labeled material transferred from the viruses to the nonradioactive second batch of bacteria was *DNA*, and *not any* protein. The Hershey-Chase experiment provided the first direct evidence that DNA is indeed the heritable factor of the gene (Carter 1996; Hershey and Chase 1952).

## After Hershey and Chase, the next chase is to find the chemical structure of DNA

Through 1952 and 1953, major scientific insights into the nature of the heritable factor, the mechanism of inheritance and evolution, now recognized by science as DNA, moved with breathtaking speed—not a common scientific phenomenon up until that time. Now the race was on to unlock the secret of how the mechanism works and win the Nobel Prize. In order to do so, the structure of the mechanism would have to be deciphered in more detail. Although the chemical composition of the mechanism was known, as described above (figures 3.1, 3.2), how these chemical pieces fit together was not.

In 1952, Erwin Chargaff at Columbia University in New York demonstrated that the structure of a DNA polymer macromolecule isn't just a monotonous repetition of the same old nucleotides over and over again. Rather, the four kinds of nucleotides could follow one another in any of a very large number of possible sequences. And biological evidence was found to back up the chemical evidence. It was found that the relative abundance of the four bases was different in DNA samples from each of many different kinds of living organisms. The chemical identities of the four kinds of bases had been worked out previously. Two of them are slightly larger than the other two. The large ones (purines) are adenine and guanine. The small ones (pyrimidines) are thymine and cytosine. Now it could be (and was) envisioned that the genetic information of the heritable factor was carried in the form of a precise sequence of bases along the DNA nucleotide chain. The firewood was in place and smoking, and the heat was growing intense. It was only a matter of time before the discovery of exactly how the mechanism of DNA works would ignite.

Chargaff made another discovery in 1952, in which he used the above information as a clue. This discovery was crucial to the discovery of the structure of DNA the following year and to how that structure allows the DNA molecule to function as the precise mechanism of heredity and evolution that it does. By comparing the ratios by weight of various combinations of bases in the DNA of various organisms, Chargaff found that the ratios of adenine to thymine (A:T) and of cytosine to guanine (C:G) were constant for all DNA of all species of organisms that were looked at. In every case, these ratios were 1:1. So although the diversity of ratios of the four base types among species of living organisms was found to be great, this could not be said for the ratios of these two pairs (Stent, xv).

Rosalind Franklin was a tragic heroine in the discovery of the molecular structure of DNA, and no discussion of this momentous event should exclude her name and contribution. James Watson, an American, and Francis Crick, an Englishman, are generally credited with this discovery. James Watson was not exactly kind to her in his book *The Double Helix* (Watson 1957, 14–15, 43–46, 86–87, 95–96, 124), which was finished after her untimely death at only thirty-seven. He and Francis Crick likely would have been set back by years in their race to discover the DNA's structure with solid evidence and win the Nobel Prize had it not been for Franklin's x-ray crystallography work on DNA and consultation with them. Franklin's work provided crucial confirmation for the outside position of the phosphate groups on the DNA molecule and for the helical configuration of DNA (Klug 1968; figure 3.7). None of these contributions were given the merit they deserved in Watson's book. Had she lived, Franklin likely would have shared the Nobel Prize with Watson and Crick.

To put it succinctly, Watson and Crick were dedicated, state-of-the-art researchers who had the vantage point offered by the times. They lived at a historical moment in which they could draw on the great achievements of others described above as well as on new technology. From Franklin, they knew that the phosphate groups are on the outside of the DNA molecule. From Chargaff, they knew that the bases can occur in a huge array of different sequences along the molecule and that the ratio of A to T and C to G is always 1:1 in the DNA of all living things tested. From this knowledge base, their own analyses, and their "Tinkertoy models" of how the various molecular groups might be related, their great discovery emerged (figures 3.7, 3.8).

# How does the heritable factor work?

In their article reporting their finding in *Nature*, a British journal dedicated to important research in all areas of science from all parts of the world, Watson and Crick wrote: "It has not escaped our notice that the specific pairing we have postulated immediately suggests a possible copying mechanism for the genetic material" (Watson and Crick 1953). This sentence has been called the biggest understatement in biology. Watson and Crick were very aware at the time that their discovery had enormous meaning for heredity and evolution beyond a "pretty structure." They were referring to their discovery of how Chargaff's rule, the observation of the 1:1 ratios of the bases T:A and C:G in DNA, squares with DNA structure. Their explanation was that in the structure of DNA, T always and only pairs with A and vice versa, and G always and only pairs with C and vice versa. This generalization has become known as the base-pairing rule (see figure 3.8). In effect, they said that they recognized that this pairing rule provides the explanation for the way that the mechanism works to carry heritable (genetic) information, as well as to reproduce it by precise copying. What is the connection between the base-pairing rule and how a DNA molecule functions? How does the sequence of bases in a DNA molecule spell out a phenotypic trait in an organism? The answer to the first question has to do with DNA replication. The answer to the second lies in how the DNA sequence in a given gene is translated into a specific kind of protein molecule.

Cytosine
Guanine
Adenine
Thymine

**Figure 3.7. DNA double helix structure. The figure represents a short section of a DNA molecule. The two winding gray helical spirals represent the sugar-phosphate backbone. Each inside "step in the staircase" represents a pair of complementary bases, or a base pair. Note that cytosine only pairs with guanine, and adenine only with thymine. The number of possible unique base sequences for a complete DNA molecule, thousands of base pairs long, is immense.**

# How does DNA replicate itself?

DNA replication occurs with the help of a team of proteins, including *helicase* and *DNA polymerase*. First, DNA is uncoiled by the protein helicase. The weak bonds between the complementary bases are broken so that the two strands are separated (figure 3.8A). Then two DNA polymerase protein molecules begin their work of gliding along each of those mother strands, collecting free-roaming DNA nucleotides in the cell and lining them up so that they correspond to the base sequence of their complementary *mother strand* nucleotides (figure 3.9, 3.10, 3.11). The products of this process are two new daughter DNA molecules, each with a mother strand and a daughter strand (figure 3.11). Whenever cells divide, including those that eventually give rise to gametes, the cell division is preceded by DNA replication (as was seen in chapter 2).

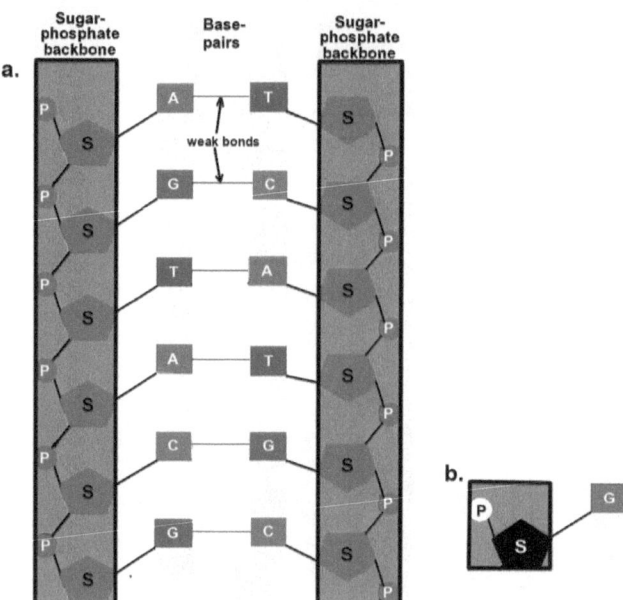

**Figure 3.8. DNA structure with double helix uncoiled. P=phosphate group. S=deoxyribose sugar. Nitrogenous bases are designated as A (adenine), C (cytosine), G (guanine), and T (thymine). A. The structure is composed of two outer sugar-phosphate backbones, with the bases inside the structure. Weak bonds between the complementary bases hold the two strands together. B. A nucleotide, the basic unit, or monomer, of a DNA molecule. There are four kinds of DNA monomer, each having one of the four types of base.**

## How is the genetic code of DNA translated into outward expression (phenotype), and where do new alleles come from?

But what have proteins got to do with phenotype? Remember that the phenotype is the outward expression of a gene in an organism. One version of a gene (allele) causes blue pea flowers, another white. One allele in the human genome causes an ear without an earlobe, and another causes an ear with an earlobe as we saw in chapter 2 (figure 2.8).

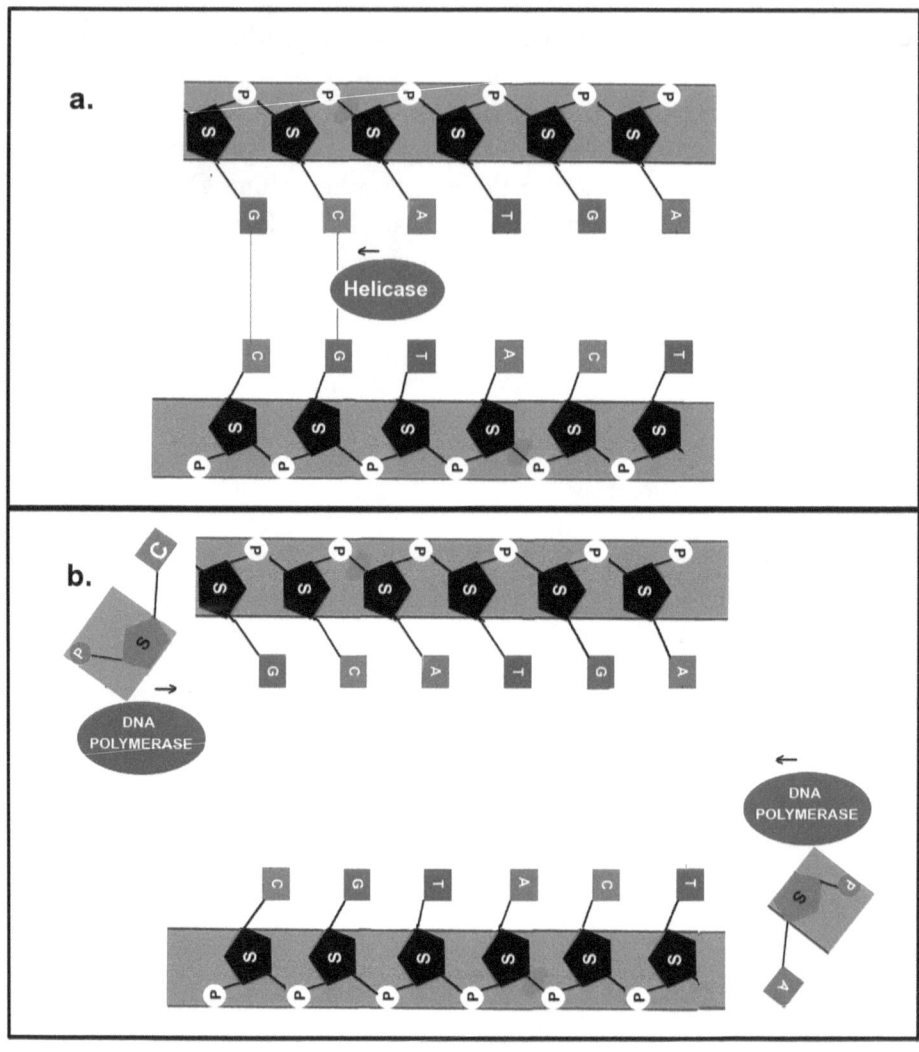

Figure 3.9. Beginning of DNA replication. A. First, helicase (protein) enzyme breaks the weak hydrogen bonds holding together the two strands of DNA by the bases. B. Next, DNA polymerase enzymes enter the scene. Each begins its glide along one of the now separated DNA strands, attracting a nucleotide (DNA monomer) with a base that is complementary to the first base as it glides past on its DNA strand. The DNA polymerase in the upper left begins by attracting a nucleotide with the base C, which is complementary to the first base it encounters in the DNA strand (G).

JOSEPH FORTIER

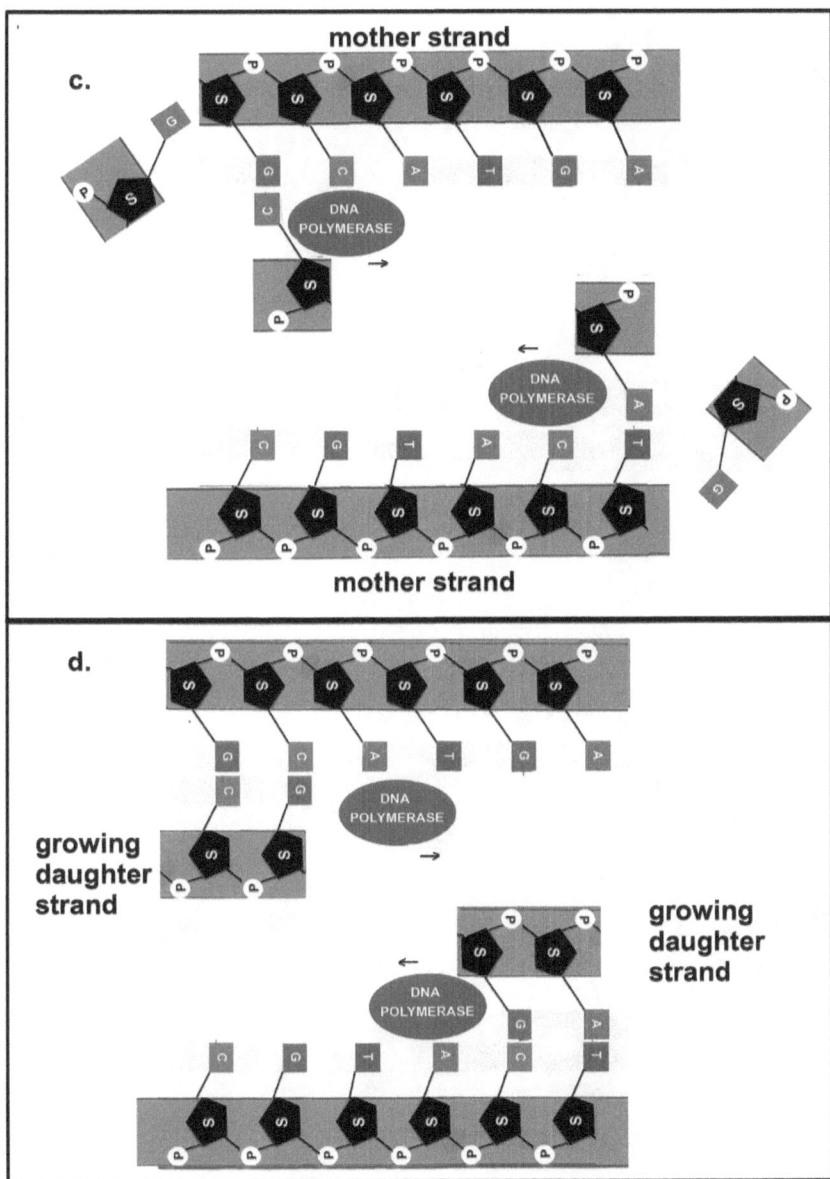

**Figure 3.10. DNA replication. C.** As each DNA polymerase molecule glides along a DNA strand, it attracts and affixes successive nucleotides with appropriately complementary bases to the bases in the mother strand. **D.** As the phosphate molecules of the newly inserted nucleotides bond to the sugars of the previously inserted nucleotides and vice versa, and as weak bonds form between complementary bases, the daughter DNA strand grows.

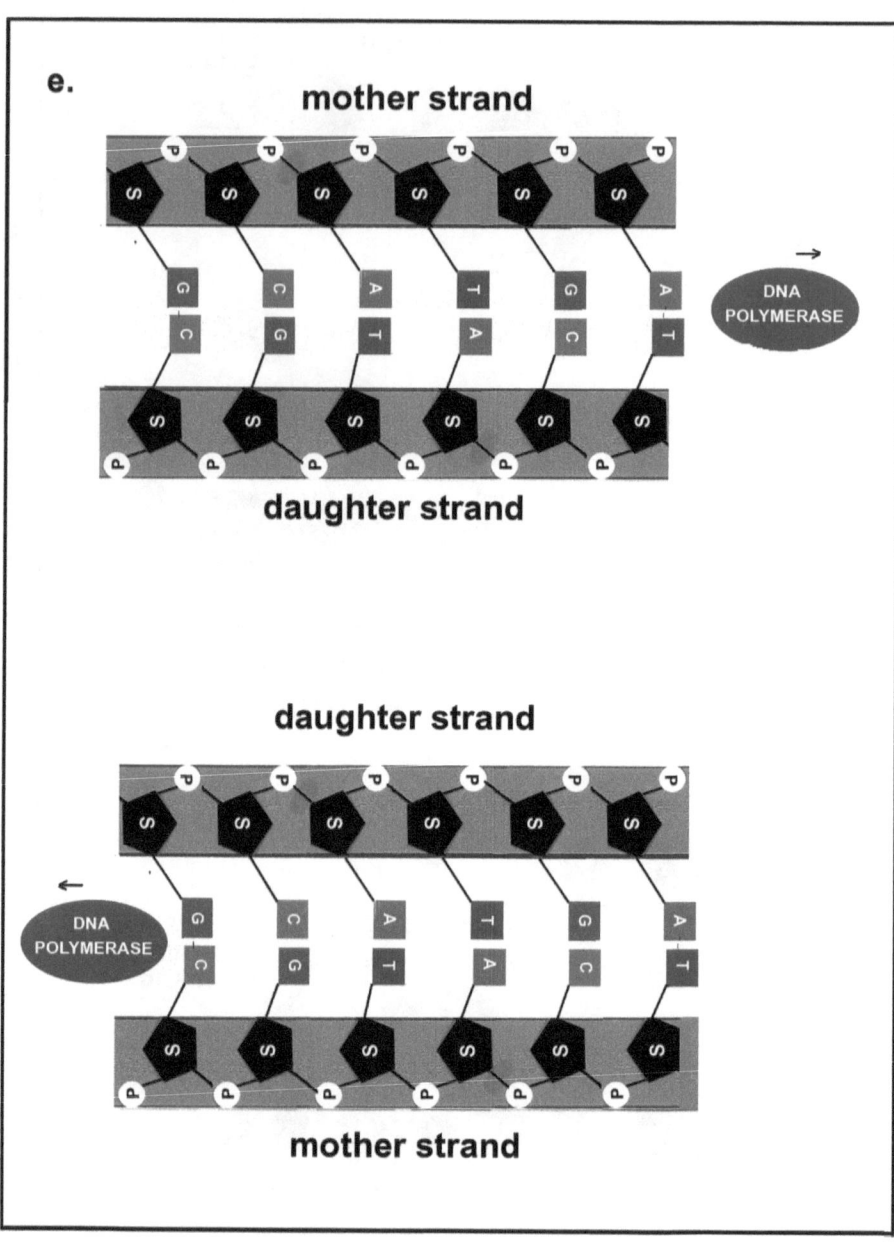

**Figure 3.11. Termination of DNA replication.** When each DNA polymerase reaches the end of the newly formed DNA molecule, it leaves the molecule. Now each of the two new DNA macromolecules consists of an old (mother) strand that was used as a template for building the new (daughter) strand, and the daughter strand itself.

JOSEPH FORTIER

It turns out that proteins have everything to do with an organism's appearance and functioning. They determine the uniqueness of an organism or group of organisms, causing them to look as they do and function as they do. There is at least one kind of protein that is responsible for every single tiny event that occurs in every cell in our bodies. Since these events are very complicated and interdependent (meiosis is just one example out of millions), there need to be (and are) veritable, swarming, myriad kinds of proteins to carry out these biochemical events. *Enzyme proteins* are chemicals that facilitate functions like growth, digestion of food, cell division, hair growth/nongrowth, and hundreds of thousands of other bodily functions. They do so by controlling the huge number of little details needed to accomplish these functions so that these little events and, thus, the big culminating events can happen in a flash rather than things taking years. Life would be difficult if it took weeks to raise an arm to scratch an itch. Thanks to enzyme proteins, things happen faster than that. *Structural proteins* are components of an organism's physical build. Examples are keratin in hair and nails, collagen in our skeletons and joints, and myosin, a key component of muscle cells. The overall functioning of these millions of kinds of proteins determines all the physical characteristics of an organism and a good many of its behavioral characteristics. In short, the kinds of proteins an organism has express the appearance and functioning of the organism. The sequence of bases in the DNA is the written music; the activity of proteins is the orchestra.

Like DNA, protein is a polymer. Unlike DNA, the monomers that comprise a protein molecule are *amino acid* molecules rather than nucleotides. Also, unlike DNA, there are twenty different kinds of amino acids rather than the four kinds of nucleotides in DNA. Again, unlike DNA, each kind of protein is contorted into a unique shape, a shape that is dictated by the particular sequence and number of amino acids in the protein polymer. By and large, it is the *specific shape* of a particular protein that allows it to carry out its particular *function* in just the way it does.

**Amino acids**

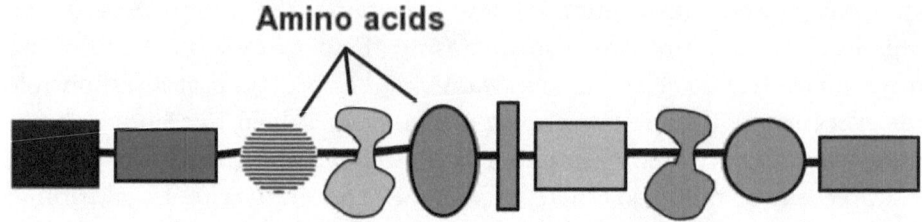

Figure 3.12. Diagram of a short section of a protein polymer. Each differently shaded and shaped figure represents a different kind of amino acid. In reality, a protein may consist of hundreds or thousands of amino acid monomers. The number, kind, and sequence of amino acids determine the precise shape that the protein polymer will fold into and, thus, how it will function.

Now what is the connection between DNA, proteins, and phenotype? It turns out that the sequence of nitrogenous bases in the DNA molecule is a language that is translated into a corresponding sequence of amino acids in a protein. A specific amino acid sequence causes a specific protein shape; a specific protein shape causes a specific activity that characterizes a specific form of life and, depending on the protein, a specific individual. Each of the effects of these protein-mediated activities constitutes an aspect of the individual's outward phenotype.

So how are proteins synthesized using information from DNA? The first part of the story is the *transcription* of the language found in the sequence of the DNA in a gene (a gene, remember, is a segment of a chromosome). The gene is transcribed into a close chemical relative of DNA: messenger RNA, or *mRNA* (messenger ribonucleic acid). The mRNA differs from DNA in three ways. First, the ribose sugar in the sugar phosphate backbone is slightly different from the deoxyribose in DNA. Second, mRNA has a base that replaces thymine (T) in DNA—uracil (U). However, uracil, like thymine, binds with adenine (A). Finally, RNA only has a single strand, rather than the double strandedness of DNA.

The first step in the transcription process requires the protein *RNA polymerase* (figure 3.13). The protein reads the base sequence on one of the two DNA strands (the template strand) and strings together an mRNA molecule with nucleotides complementary to those of the DNA template strand. That is, for every template strand nucleotide in which the base is T, the nucleotide assembled at the corresponding site on the growing

JOSEPH FORTIER

RNA molecule will have the base A, which is always complementary to T. (This is one of the base-pairing rules. The other one is that G and C are always complementary to one another.) For every DNA A, the RNA polymerase will insert an RNA nucleotide with the base U. There are a lot of nucleotides floating around inside the cell. The RNA polymerase protein simply attracts the appropriate complementary mRNA nucleotide for the DNA template strand nucleotide it is presently reading and fits it into the correct position in the growing RNA molecule, which remains connected to the RNA polymerase protein (figure 3.13). Once a gene has been transcribed into a complete mRNA molecule, that mRNA molecule detaches from the RNA polymerase protein and migrates away from the cell nucleus, out into the cytoplasm, or outer region of the cell (figure 3.13E). There, in a beautifully complex and elegant process, mRNA acts as a template for the stringing together of just the right order of amino acids. So a protein is built, with amino acids, often hundreds of them, assembled in a sequence corresponding to the sequence of mRNA bases, often thousands of them (figure 3.12).

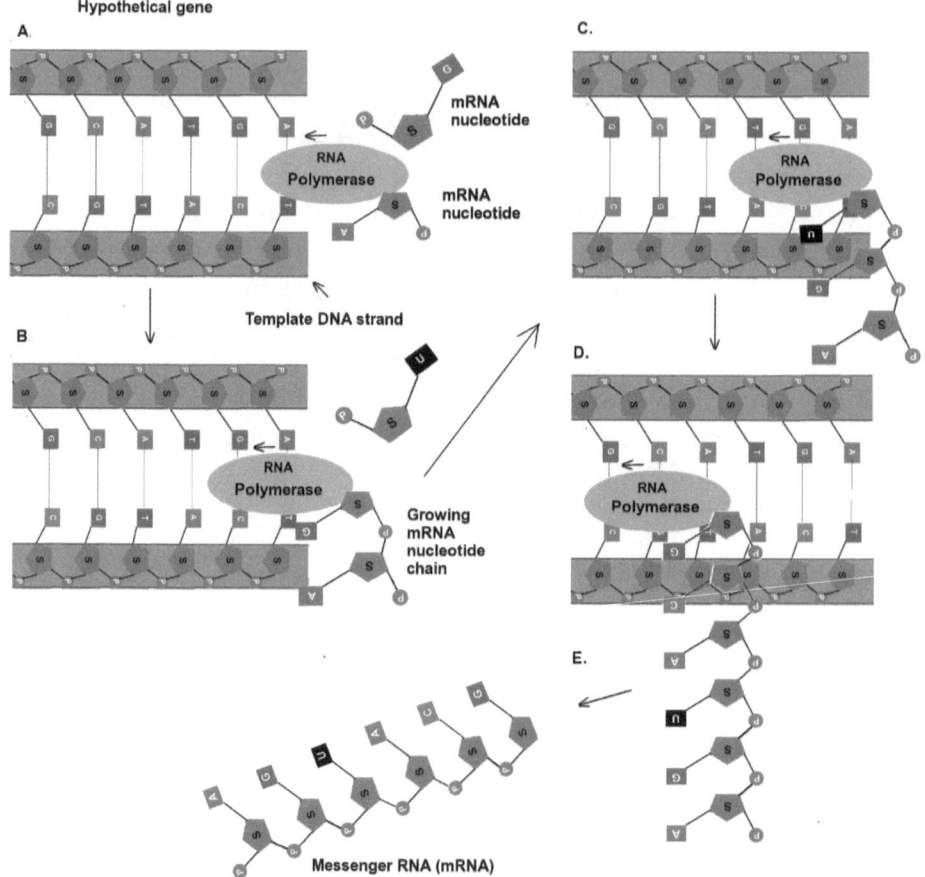

Figure 3.13. Transcription of a mRNA molecule from DNA in a gene. For clarity, the hypothetical gene is ridiculously small—most genes are thousands of nucleotides long. B-C. As the mRNA polymerase moves from nucleotide to nucleotide of the DNA template strand, it gathers base sequence information a base at a time, and transcribes it into a complementary strand of mRNA. The mRNA nucleotides are gathered from the cell. D. RNA polymerase reaches the end of the gene. E. The completed mRNA strand is released from the mRNA polymerase molecule.

JOSEPH FORTIER

To make a long story short, the DNA base sequence of a gene is the remote determinant of a specific protein's amino acid sequence. The mRNA, the near determinant, mediates the process, getting its orders from the DNA in a gene (really *being* the orders it gets) and translating those orders into an amino acid sequence (translation). After the protein curls and rolls into its characteristic shape, which is determined by its amino acid sequence, it receives its function from that form and confers a specific character to the organism by performing a specific task in a certain way—often in concert with the work of other proteins.

## *Where do new alleles come from?*

In chapter 2, we saw that Mendel worked with versions of the heritable factor of a gene called alleles. We also saw that Thomas Morgan and his graduate student, A. H. Sturtevant, discovered that new alleles in their fruit flies could be generated with x-ray irradiation. What is the mechanism by which new alleles are generated in a population? Thanks to the discovery of DNA structure, we now know the answer. Additionally, knowing the basis in DNA replication for generating new alleles has spawned such biotechnical industries as genetic counseling, in which specialists can screen a couple for possible genetic diseases caused by unhealthy mutant alleles that may be passed to their potential offspring.

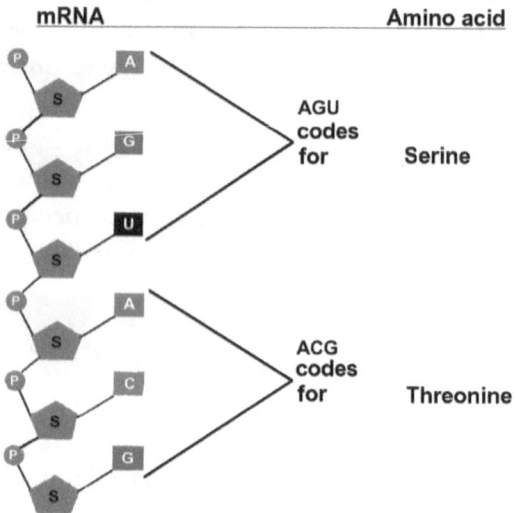

mRNA          Amino acid

AGU codes for    Serine

ACG codes for    Threonine

**Figure 3.14. Translation and the genetic code. Each sequence of three mRNA bases (a codon) codes for a specific amino acid. There are 20 kinds of amino acids, two of which are serine and threonine. Thus, the base sequence of the DNA of a gene codes for a specific protein polymer (monomers are amino acids). That code is transcribed into the mRNA code and then translated from the mRNA sequence into an amino acid sequence. For the genetic code, see reference (Kimball 2008).**

When DNA replicates itself with the help of DNA polymerase, sometimes the polymerase protein makes a mistake and inserts a nucleotide that is not complementary to the nucleotide in the matching mother strand, into a site in the sequence of the new, growing sister strand. This sort of mistake results in what is called a "substitution mutation" (Fig. 3.15). When this happens, there are three possible outcomes. One is that nothing happens (figure 3.15, middle). When nothing happens, here is the scenario. An incorrect nucleotide is substituted for the correct one at a given site on the DNA of a gene. The DNA is transcribed into an mRNA molecule. The resulting mistaken mRNA codon, which is the product of the original DNA mistake, codes for *the same amino acid* in the protein as would have been coded for had the mistake not been made. Thus, there will be no change in the resulting protein and no phenotypic effect. Second and third outcomes arise from a different amino acid sequence than in the normal protein product, resulting in a differently functioning protein (figure 3.15, bottom). These outcomes are (1) death/harm or (2) increased fitness.

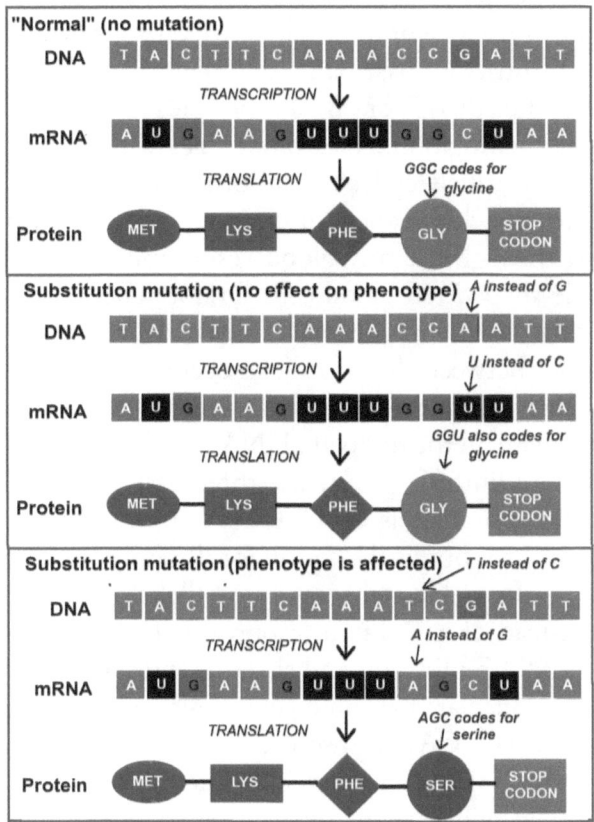

**Figure 3.15. Substitution mutations.** Top box shows a normal transcription and translation process. Middle box shows a mistaken substitution of adenine (A) for guanine (G) during DNA replication. Since the resulting mRNA codon (GGU) codes for the amino acid glycine as does normal mRNA with codon GGC (top box), there is no outward effect—the mutation remains hidden. Bottom box shows a mistaken substitution of thymine (T) for cytosine (C) in DNA replication resulting in codon AGC instead of the normal codon GGC. AGC codes for the amino acid serine instead of glycine. Thus, the mutation is expressed as a different protein that will function differently from the normal protein.

# Sickle-cell anemia: a case of natural selection for a sometimes lethal, sometimes benefical allele

Sickle-cell anemia is a genetic disease that originated in a part of the world where malaria is a common disease. We're dealing here with two different diseases. One is sickle-cell anemia, a genetically caused condition, and the other is malaria, a parasitic disease. The gene that codes for hemoglobin, the protein in red blood cells that carries oxygen in the blood, has two alleles. Both of these alleles are versions of a gene that is 861 nucleotides long (thus, 861 base pairs [bp] long). Each protein component of hemoglobin is thus 861/3, or 287 amino acids long. One of these alleles codes for normal hemoglobin. Red blood cells with normal hemoglobin are round. The other allele codes for abnormal hemoglobin, which causes red blood cells to be sickle shaped, or crescent shaped. Normal-shaped red blood cells flow through the tiny blood vessels called capillaries smoothly and easily. Sickle-shaped red blood cells do not. They clog the capillaries. People with sickle-cell anemia often die very early in life.

The difference between the two alleles occurs at only one base-pair site (figure 15). At this site on the allele that codes for non–sickle cell, normal hemoglobin, the base on the template DNA strand is T. The bases on either side of that site are both C. Thus, the three-base sequence at this point on the non–sickle cell allele is CTC. CTC is transcribed to GAG in the mRNA, which, it turns out, is the codon for the amino acid called glutamic acid. However, in the sickle-cell version, or allele, of the gene, there is an A at that site instead of a T due to a previous DNA polymerase-reading mistake during DNA replication and sister-strand synthesis (figure 15). So the three-base sequence at that same genetic site on the sickle-cell allele is CAC. instead of CTC. CAC transcribes to GUG on the mRNA, which in turn is the codon for the amino acid valine. So the transcribed mutant hemoglobin protein will have a valine rather than a glutamic acid at the resulting site, which causes the protein to be shaped differently, which causes the red blood cell to be shaped differently, which causes the genetic disease. Now the header to this section says that this allele, caused by the genetic substitution mutation described above, is sometimes beneficial or adaptive. How can that be?

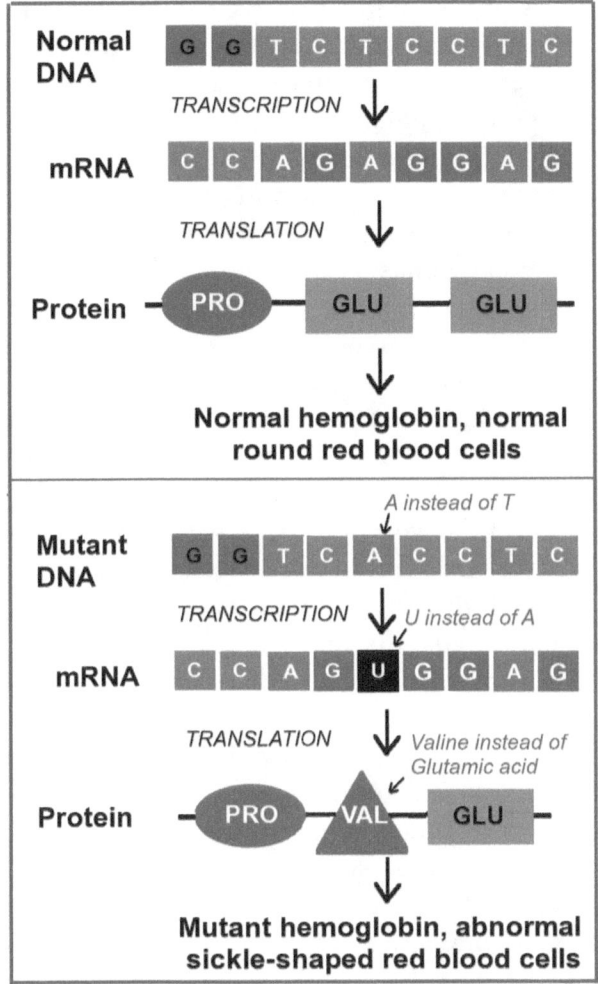

**Figure 3.16. Substitution mutation and sickle-cell anemia.** Top box shows a nine-base segment of DNA of a gene that codes for normal hemoglobin protein. Middle mRNA codon (GAG) codes for the amino acid glutamic acid. Bottom box shows the same DNA locus in the gene, except that it shows the effects of one base substitution mutation, causing the entire 861 nucleotide-long gene to code for mutant sickle-cell hemoglobin. Note that the fifth base from the left is A instead of T (a substitution mutation). A is transcribed as mRNA U, and since the codon GUG is translated as the amino acid valine (instead of normal GAG translating as glutamic acid), the 287 amino acid-long protein has one incorrect amino acid (valine). This one mistake causes a dramatic change in the shape and, thus, functioning of the protein. The red blood cell with this mutated hemoglobin protein is sickle-shaped if the genotype is homozygous. It is intermediate in shape between round and sickle-like if the genotype is heterozygous.

The reason lies in the dominance/recessive relationship in how these alleles influence each other's expression and on the presence of the deadly disease malaria, where this mutation originated in the human population. When the genotype is a homozygous-mutant type (both alleles code for sickle-cell anemia at that gene locus in the homologous chromosomes), the phenotypic expression will be harmful or lethal. When the genotype is homozygous-normal type (both alleles code for normal hemoglobin at that gene locus in the homologous chromosomes), the phenotypic expression will of course be normal. *Here is the rub*: when *both alleles* are present at that gene locus (heterozygous genotype), the phenotype will be an intermediate condition; *both alleles will be partly expressed.* The resulting phenotype will be intermediate between round red blood cell and sickled red blood cell. This phenotype results in *only a mild form* of anemia that allows the person to live a normal life and be very *resistant to malaria,* which is a major child killer in many areas of the world. It can be seen how this mutation, which appeared as a mistake, had survival advantage in areas heavily infested with malaria so people with the heterozygous genotype had a strongly adaptive phenotype—even more adaptive than the normal-hemoglobin phenotype. They could live past childhood without dying of either malaria or sickle-cell anemia to reproduce themselves. The sickle-celled mutant allele was selected for by natural environmental conditions (natural selection) in these parts of the world where malaria is common.

In conclusion, the functioning of the DNA (heritable factor) is to be found in its chemical structure (form). DNA's self-replicating, information-storage, and information-transmission functions comprise the mechanism of inheritance and evolution found in its structure, including the specific sequence of its four kinds of nucleotide monomers and their complementary base-pairing in the double helix of DNA. In the organism, DNA's stored and transmitted information is used for everything biological that expresses the organism's being—via the activity of the proteins. A protein's amino acid sequence—and thus, form and functioning—arises from the information encoded in the DNA. It is not hard to understand why the former director of the National Human Genome Research Institute Dr. Francis Collins felt inspired to name his book *The Language of God* (Collins 2006). The heritable factor—with its intricate yet simple, compact informational and replication mechanism that allows for the passing down of inherited information from ancestors to descendants, including modifications of that information via mutations—is indeed full of grandeur in all its simplicity, to use a theme from the last sentence of *Origin of Species* (Darwin 1859). Of course, Darwin was describing the "grandeur from simple beginnings" of descent with modification, or evolution, caused by environmental pressures over time on inherited characteristics in populations of living organisms (natural selection). Had he known of the existence of the mechanism of heritability with modification, the DNA of the gene, and its grand, simple structure, I wonder if he would have agreed with the title of Dr. Collins's book.

# Summary

☐ The discovery of the mechanism of evolution and heredity, or heritable factor, was a collective scientific enterprise in which the work of individual scientists built on the work of others.

☐ Scientists remained skeptical, as is their nature, about new discoveries that suggested that DNA is the mechanism of heredity and evolution until it was demonstrated beyond doubt by the Hershey and Chase experiment.

☐ DNA is a polymer macromolecule in which the molecular subunits, or monomers, are nucleotides. A nucleotide is in turn composed of one small molecular subunit of each of the following: phosphate, pentose sugar, and nitrogenous base. There are four kinds of nitrogenous bases.

☐ The sequence of bases along the DNA molecule comprises a code, or language, that carries all the heritable information in an organism.

☐ The DNA code can be replicated and perfectly preserved when DNA is replicated. This feature of the DNA macromolecule provides the mechanism for inheritance and evolution.

☐ The genotype of an organism is the sum of the information carried in the organism's DNA molecules.

☐ The genotype becomes expressed as the organism's phenotype when the information in the DNA molecules is transcribed into the intermediary form of messenger RNA molecules (mRNA) and the mRNA information is translated into chains of amino acid molecules. These chains of amino acid molecules are protein polymer macromolecules that each have a specific function in an organism. The exact sequence of amino acids in each protein determines the exact way each protein performs its function, which in turn determines the outwardly expressed biological uniqueness of particular groups of organisms and even individual organisms.

☐ Mutations sometimes occur in which a mistake is made in the replication of a DNA macromolecule, and the mistake is passed on to gametes. Some of these mutations are beneficial, may be selected for under natural environmental conditions, and become new alleles in a population of organisms.

# Glossary

**adenine (A)**. One of four kinds of base molecules found in DNA and RNA (figure 8).

**amino acid**. One of twenty small molecular monomers of proteins molecules (figure 11).

**bacteria**. A one-celled living organism without a separate membrane enclosing the genetic material in the cell.

**bacteriophage**. A type of virus that attacks bacterial cells (figures 4, 5).

**base**. See **nitrogenous base**.

**codon**. In mRNA molecules, a sequence of three RNA bases that codes for one of twenty kinds of amino acids. Thus, a sequence of many codons in an mRNA molecule provides the information for the cell to string together amino acids in the correct sequence to manufacture a particular kind of protein (figure 13).

**complementary bases**. Two bases in DNA or RNA that form weak hydrogen bonds with each other (figure 8a). Adenine (A) and thymine (T) are always complementary to each other in DNA. Adenine and uracil (U) are always complementary to each other in RNA. Cytosine and guanine are always complementary to each other in both DNA and RNA.

**cytosine (C)**. One of four kinds of base molecules found in DNA and RNA (figure 8).

**daughter strand**. During DNA replication, the new DNA strand that is built by a DNA polymerase molecule and becomes part of the new DNA molecule along with the mother strand (figure 9d–e).

**denature**. The process of changing the chemical structure of a molecule often by heat.

**deoxyribose.** A sugar molecule component of DNA consisting of five carbon atoms and three oxygen atoms. Together with phosphate, deoxyribose is a major component of the sugar-phosphate backbone in DNA.

**DNA (deoxyribonucleic acid).** A large polymer macromolecule consisting of a double helix of nucleotide subunits (figures 7, 8).

**DNA polymerase.** An enzyme that is active during DNA replication that facilitates the formation of a daughter strand of DNA, using the mother strand as a template for correct ordering of the base sequence in the daughter strand (figure 9b–e).

**double helix.** A conformation in which two linear objects wind around one another several times (figure 7).

**enzyme.** A protein that is used to facilitate, generally by enormously speeding up, a chemical reaction.

**guanine (G).** One of four kinds of base molecules found in DNA and RNA (figure 8).

**helicase.** An enzyme that is active during DNA replication that facilitates the unwinding of the double helix DNA molecule, as well as breaking apart the hydrogen bonds that hold the complementary bases together (figure 9a).

**hemoglobin.** A protein found in red blood cells that is composed of four protein subunits. Each protein subunit is a macromolecule polymer composed of many amino acid molecules.

**macromolecule.** Any large molecule consisting of hundreds or thousands of atoms.

**messenger RNA (mRNA).** A macromolecule polymer composed of nucleotides with ribose sugar molecules, rather than deoxyribose sugar molecules such as in DNA, and never coiling into a double helix as DNA does. mRNA molecules are produced by the process of transcription and, in turn, are used as templates in the cell for manufacturing proteins by facilitating the correct sequence of protein monomers (amino acids) (figure 13).

**monomer.** Small molecular subunits of larger molecules called polymers (figure 1).

**mother strand.** During DNA replication, after the helicase enzyme breaks the hydrogen bonds holding the complementary bases of the DNA together, the DNA strand that is used as a template for forming the new daughter strand (figure 9c–e).

**mutation.** A mistake that is sometimes made in the base sequence of a sister DNA strand while the strand is being formed by a DNA polymerase molecule. Some mutations are harmful since they result in a dysfunctional or nonfunctional protein molecule. Others are beneficial when they result in an altered protein molecule that is still functional although in a slightly different way (figure 15).

**nitrogenous base.** One of the five base molecules that are components of a nucleotide of DNA and RNA. The nitrogenous base molecules found in DNA are adenine (A), cytosine (C), guanine (G), and thymine (T). The nitrogenous base molecules found in RNA are adenine (A), cytosine (C), guanine (G), and uracil (U) (figure 8).

**nucleotide.** A molecular subunit or monomer of DNA or RNA. A nucleotide consists of a phosphate molecule, a pentose sugar molecule, and a base molecule (figure 8b).

**pellet.** The fraction of a solution after centrifuging that remains at the bottom of a test tube, as a solid (figure 6).

**phosphate.** A small molecule composed of an atom of phosphorus and four oxygen atoms. Together with deoxyribose sugar, phosphate is a major component of the sugar-phosphate backbone in DNA (figure 8).

**polymer.** A large molecule consisting of hundreds or thousands of atoms and consisting of chains or nets of smaller molecular subunits (figure 1).

**polysaccharide**. A starch or cellulose molecule; a polymer consisting of many sugar molecules as the monomers.

**RNA polymerase**. The enzyme that facilitates the synthesis of RNA (figure 12).

**sickle-cell anemia**. A genetic disease caused by a substitution mutation that has become part of the human genetic makeup in some individuals (figures 14, 15). When the mutation is inherited from only one parent, it may confer resistance to becoming seriously ill from malaria. If it is inherited from both parents, it results in a serious genetic disease.

**structural protein**. A protein that is a component of a cell, giving the cell structure or shape or, in the case of muscle cells, causing cells to contract or lengthen.

**substitution mutations**. A mistake in DNA replication in which the wrong base is put in a particular place in the sequence of the growing daughter DNA strand. When the mutated daughter DNA strand is used as a template for mRNA synthesis, the mistake is transcribed into the new mRNA. During translation from the mRNA code of base sequences to the amino acid sequences in a protein being synthesized, the protein synthesis may or may not result in a mutated, dysfunctional protein (figure 14).

**sugar-phosphate backbone**. In a DNA or RNA molecule, the chain consisting of phosphate and sugar molecules (figures 7, 8a).

**supernatant**. The fraction of a solution after centrifuging that remains liquid and is not found at the bottom of a test tube (figure 6).

**T2 phage**. A bacteriophage that attacks the bacteria *Escherichia coli* (*E. coli*).

**thymine (T)**. One of four kinds of base molecules found in DNA (figure 8).

**transcription**. The process whereby an RNA molecule is produced in which the enzyme RNA polymerase uses a strand of DNA as the template for correct sequencing of the nucleotide bases in the RNA molecule (figure 12).

**transformation.** The process by which bacteria of a given strain absorb the genetic material from bacteria of another strain, and that absorbed genetic material expresses itself, transforming the recipient strain into the donor strain (figure 3).

**translation.** The process in which a protein is built as an amino acid chain, using the codon sequence of an mRNA molecule as the template for correct ordering sequence of the the amino acids in the chain (figure 13).

**uracil (U).** One of four kinds of base molecules found in RNA (Fig. 12).

**virus.** A small nonliving particle composed of protein and either DNA or RNA, which must inject its DNA or RNA into a living cell, use the protein machinery of the living cell to reproduce new virus particles, and usually kill the cell in the process (figures 4, 5).

# References

Bookrags. 1999. "Frederick Griffith Biography." Last modified 2011. http://www.bookrags.com/biography/frederick-griffith-wob/

Carter J. S. 1996. "DNA Structure and Function." Last modified November 2004. http://biology.clc.uc.edu/courses/bio104/dna.htm

Collins F. S. 2006. *The Language of God: A scientist Presents Evidence for Belief.* New York: Free Press.

Darwin C. 1859. *On the Origin of Species by Means of Natural Selection.* London: John Murray.

Henig R. M. 2000. *The Monk in the Garden.* New York: Houghton Mifflin Co.

Hershey A. D. and M. Chase. 1952. "Independent functions of viral protein and nucleic acid in growth of bacteriophage." *Journal of General Physiology* 36: 39-56

Kimball. JW. 2008. "The Genetic Code." Last modified April 2010. http://users.rcn.com/jkimball.ma.ultranet/BiologyPages/C/Codons.html.

Klug A. 1968. "Rosalind Franklin and the Discovery of the Structure of DNA. *Nature* 219: 808-844.

Medawar P. B. 1980. "Lucky Jim." In *The Double Helix*, edited by G. S. Stent. New York: W.W. Norton & Co.

Biotechnology Industry Organization. 2000. "The Search for DNA—the Birth of Molecular Biology." Last updated 2009. http://www.accessexcellence.org/RC/AB/BC/Search_for_DNA.php.

Stent G. S. ed. 1980. *The Double Helix: A Personal Account of the Discovery of the Structure of DNA.* New York: W. W. Norton & Company, Inc.

Watson J. D. and F. H. C. Crick. 1953. "A Structure for Deoxyribose Nucleic Acid." *Nature* 171: 737-8.

Watson J. D. 1957. The Double Helix: A Personal Account of the Discovery of the Structure of DNA. In *The Double Helix: A Personal Account of the Discovery of the Structure of DNA*, edited by G. S. Stent. New York: W. W. Norton & Company, Inc.

# CHAPTER 4

# What is the Evidence that
# Evolution has Happened and is Happening?

IN CHAPTER 2, we saw that the scientific method is a tool for estimating the reliability and consistency of that which we can sense and analyze rationally. We saw how Charles Darwin, the founder of the modern theory of biological evolution, applied the scientific method to arrive at this theory. Furthermore, we distinguished between hypothesis and theory and saw that a hypothesis only becomes a theory after the scientific community has rigorously challenged it with repeated testing from many angles over time and has found it to be reliable and consistent. Thus, a scientific theory is a hypothesis that has become accepted by the community as tantamount to fact.

Nonetheless, scientists, being the skeptics they are, continue to put theories to the test. In order for Darwin's hypothesis of evolution to receive the credibility it needed in order to become recognized as a theory, the issue of mechanism had to be addressed credibly. As we saw, Gregor Mendel's work provided the evidence for the existence of this mechanism (the smoking gun). Mendel's work was subsequently corroborated by the discovery of chromosomal behavior during meiosis, the process by which gametes, the eggs and sperms, are produced. Finally, in chapter 3, we saw that the heritable factor, or mechanism for evolution, was identified as DNA (the bullet). We looked at how the mechanism works to store, reproduce, shuffle, and transmit information through generations.

So far, so good. But what is the evidence that descent with modification by natural selection *has really happened*? How well does this evidence hold up to the standards of reliability and consistency in supporting descent with modification by natural selection? In order to address these questions, in this chapter we'll look at the following kinds of evidence:

1. Fossil evidence and age of rock at various layers in the earth's crust and methods for measuring rock age.
2. Morphological evidence for common ancestry in diverse living creatures.

3. Biochemical evidence for common ancestry in diverse living creatures, such as similarity in proteins and DNA in diverse living creatures.
4. Embryonic evidence for common ancestry in diverse living organisms, such as similar structures in embryos of diverse living creatures, even when those structures morph into totally different structures among those creatures.
5. Evidence for actual speciation. Speciation is the process by which new species come into existence, descending from older species. A *species* is the basic category for a group of organisms, commonly defined as all organisms capable of interbreeding and producing fertile offspring. This definition is adequate in many cases for sexually reproducing organisms. In other cases, similarities and differences in DNA sequences or morphology are used to define and distinguish species (Futuyma 2005, 363–365; Lincoln et al. 1993).

## 1. Fossil evidence and age of rock

What is a fossil? How are fossils formed? What about a fossil makes it a credible piece of evidence for evolution?

*What are fossils, and how are they formed?*

*Fossils* are impressions left in rock of a formerly living organism, a piece of such an organism, or other evidence of its existence, such as a footprint or burrow. There are over twenty ways in which dead organisms may become fossilized (Canadian Fossil Discovery Centre 2009). The hard structures of organisms are those most commonly preserved by fossilization. In animals, these structures include bones; outer hard parts of insects, spiders, and crustaceans; and shells. In plants, they include pollen, wood, leaves, and flower parts. The type of rock in which these fossilized structures are most often found is *sedimentary*, meaning that fossils of dead organisms were formed when clay, silt, or sand sediment, along with the dead organism, settled to the bottom of a body of water, usually in what was shallow water with high dissolved mineral content and low oxygen content. Over a long period of time, as such sediments are buried under the weight of other such sediments, they become compacted. Under these conditions, the fossilization process called *permineralization* occurs, in which minerals in the submerged sediment particles slowly dissolve, replacing any remaining

JOSEPH FORTIER

dead organic matter during the permineralization process (Donovan 1991; Grimaldi and Engels 2005). A famous site for finding examples of insects that have been fossilized this way is the Barstow Formation in the Calico Mountains in southeastern California, USA (Grimaldi and Engels 2005, 45–46). As is most often the case with permineralized fossils, these fossils were found in what had been a shallow water body (lake in this case) with high mineral content. Another famous example of such a site is at Burgess Pass in a mountainous area in British Columbia, Canada, near the town of Field in the Yoho National Park (MacRae 2007; Matthews 2005). The type of sedimentary rock, *shale*, in which these fossils were discovered, is formed when the sediment in which the original forms of life were buried is composed of clay particles. What is fascinating about this site is that in many cases, nothing closely resembling these forms, the external morphologies of which have been preserved in fine detail, exists today. The rock in which these fossils were found is 505 million years old (MacRae 2007).

At one time, both of these areas, now in mountainous places, had been underwater. Through millions of years, geological events in the earth's crust caused low-lying, even below-sea-level areas on earth's surface to be uplifted as mountains were formed. The science of mountain formation is called *orogeny* (Matthews 2005).

Sometimes the fossil is a small organism such as an insect or bit of pollen in a piece of *amber*. Amber is mineralized plant resin or sap (Grimaldi and Engels 2005). Some fossils are the compressed black carbon remains of an organism, such as in many plant fossils and some insect fossils (Donovan 1991; Grimaldi and Engels 2005; Canadian Fossil Discovery Centre 2009). The process by which the latter is formed is called *carbonization*.

*What about a fossil makes it a credible piece of evidence for evolution?*

In order to clearly answer this question, we need to first take another look at sedimentary rock, other sorts of rock, and rock layers. Besides sedimentary rock, there are two other major types: igneous and metamorphic. Igneous rock is molten lava that has solidified. The lava comes from volcanoes or midoceanic ridges where new crust is being formed. Fossils are never found in igneous rock for obvious reasons. Even if sedimentary rock had subducted (sunk deep into the earth), eventually becoming molten, spewed forth, and thus converted into igneous rock, it would be hard to conceive of fossils enduring such events. Metamorphic rock originates from igneous and/or sedimentary rock subjected to high temperature and pressure. Any fossils present in metamorphic rock are almost always contorted and broken beyond recognition (Futuyma 2005).

James Watt, the fellow who refined the steam engine in the eighteenth century so that it became the major engine of the Industrial Revolution at that time, is indirectly responsible for the discovery that bedrock occurs in layers and that these layers are characterized by distinctive kinds of fossils only found in each layer. William Smith, another Englishman, oversaw the digging of the Somerset Canal in southern England using the new powerful, earth-moving steam-powered machinery. It was he who first observed that rock layers often appear in a regular order. He wrote, "Each stratum contained organized fossils peculiar to itself, and might, in cases otherwise doubtful, be recognized and discriminated from others like it, but in a different part of the series, by examination of them" (Miller 1999; Waggoner 1996).

Before Darwin's time, the *geological timescale* in current use by geologists and other scientists was constructed (figure 1). This chart of the ages of rock strata is an elaboration on Smith's discovery in southern England. The chart is based on extensive excavation in various parts of the world and on fossil findings. The numbers for the ages of the rock strata weren't added until later, as we'll see below. The chart remains in practical use today for such geological purposes as oil and coal exploration. As can be surmised from the chart, these rock strata can be identified from the fossilized organisms peculiar to each stratum. All over the face of the earth, in any given layer of sedimentary rock, the same kinds of fossils are found in only that layer, which correspond in age to sedimentary rock of the same age in other parts of the world, with the same fossils peculiar to that layer and that age. Successively deeper layers of rock in the earth's crust correspond with increasingly older rock and with fossils increasingly different from creatures we see today (Miller 1999; Futuyma 2005).

So how did scientists come by the numbers for the age of the rock strata in the chart? After all, even the famous physicist Lord Kelvin, in 1862, placed the age of the earth at only between 24 million and 400 million years. Now it's been determined to be about 4.5 *billion* years old (Dalrymple 1991). That's 4.5 thousand million years. How did they arrive at not only this number, but those also large numbers in figure 4.1?

In a word (or two), radiometric dating.

*Age of rock at various layers in the*
*earth's crust and methods for measuring rock age*

*Radiometric dating* only measures the age of igneous rock. Fortunately, in rock strata, the sedimentary rock substrata in which fossils occur are bounded above and below by igneous rock. By getting a read on the age of these igneous layers, scientists can get time boundaries around which the sedimentary rock was formed and, thus, an estimation of when the organism represented by the fossil lived and died. Radiometric dating of rock works because igneous rock, at the time it solidifies from molten magma, always has traces of certain radioactive substances in it. These unstable radioactive substances, over time, decay into peculiar varieties (*isotopes*) of stable substances. For example, rock age is commonly calculated using *uranium 235* (U-235). Over a lot of time, U-235 decays to stable *lead 207* isotope (Pb-207). Since U-235 decays at a constant rate into Pb-207, scientists have been able to calculate the half-life of U-235. The half-life of a radioactive substance is the amount of time it takes half of the substance to decay into the stable isotope. The half-life of U-235 is 700 million years. This means that it takes any given amount of U-235 700 million years for half of it to decay into Pb-207 (Miller 1999; Futuyma 2005). So say you found an igneous rock and you bet your friend $100 that the rock is younger than when the dinosaurs were around. First thing you do is you take a gander at the geological timescale, and then you realize that you just bet that the rock is less than 65 million years old. Next thing you do is to find a competent geologist with the equipment to measure the *ratio* in the rock of percent U-235/percent Pb-207. Say (s)he finds that the ratio is 50/50. This would tell you that half of the U-235 has decayed into Pb-207. Since you know that the half-life of U-235 is 700 million years, you know that this rock is 700 million years old.

| Era | Period | Millions of years from start of period to present | Major events |
|---|---|---|---|
| Cenozoic (Recent) | Quaternary | 1.8 | Repeated glaciations, extinctions of many large mammals, appearance of modern humans |
| Cenozoic (Recent) | Tertiary | 65 | Climate cooling, drying; radiation of mammals, birds, flowering plants, pollinating insects |
| Mesozoic | Cretaceous | 145 | Dinosaur diversity, diversification of flowering plants, mammals, pollinating insects. Mass extinction at end of period |
| Mesozoic | Jurassic | 200 | Dinosaur diversity, first birds, first flowering plants |
| Mesozoic | Triassic | 251 | First dinosaurs, first mammals |
| Paleozoic | Permian | 299 | Diversification of insects, reptiles diversify, over 90% of marine species go extinct at end of period |
| Paleozoic | Carboniferous | 359 | Large early "coal forests," first reptiles |
| Paleozoic | Devonian | 416 | First amphibians, ferns, seed plants; "age of fishes" |
| Paleozoic | Silurian | 444 | First jawed fish, first land plants, insects |
| Paleozoic | Ordovician | 488 | Diversification of invertebrates; jawless fish |
| Paleozoic | Cambrian | 542 | First appearance of most animal diversity |
| Protero-zoic | | 2500 | Earliest non-bacterial life, trace fossils of soft-bodied animals |
| Archean | | No defined lower limit | First fossil evidence of bacterial life (3500 mya). Evidence of increasing oxygen in atmosphere |

**Figure 4.1. The geological timescale.**

You've just found out that you lost your bet. In fact, if you check your geological timescale, you'd find that this happens to be about when the very first soft-bodied animal forms were crawling over the muck on the sea bottom. But say you got lucky and found that the U-235/Pb-207 ratio in your rock was 4/49. Now what? You want to find the fraction of 700 million years that 4/49 is. Since 50/50 (= 1) = 700 million years, you just

JOSEPH FORTIER

multiply 4/49 by 700 million. Since you come up with 57,142.857 years, you win the bet in this case.

In fact, the late eighteenth-century/early nineteenth-century people who developed the geological timescale didn't know about Darwin, natural selection, and descent with modification since *Origin of Species* wasn't published until 1859. They only charted what they observed. Their observations about fossils did tell them that extinction happened and clued in at least some of them that new forms of life emerged over time. No one knew *how much time* was involved. We now know, thanks to these radiometric methods that were developed in the twentieth century, that the largely extinct forms of life present in these layers of rock correspond to time periods that can be quite accurately measured and that some of them are much older than even Lord Kelvin and William Smith guessed. In addition, we know from the extensive fossil exploration that has occurred in the intervening two hundred years that there is abundant fossil evidence for forms of life that are consistently *only found* in certain rock strata around the world and *never in older deeper strata*. Strong evidence for emergence of new species. Strong evidence that evolution has in fact happened over vast amounts of time. This is just what we'd expect if the mechanism for evolution is the DNA of the gene and if the genes in populations of organisms change in commonness/rarity over many, many generations in response to the subtleties in environmental pressures and changes, as well as the relatively rare emergence, on a per-generation basis, of chance beneficial mutations.

*Is there evidence of descent with modification in the fossil record?*

By itself, the appearance of new forms of life at various times in the fossil record doesn't clinch the fossil record as evidence for descent with modification. In order for the fossil record to be persuasive evidence for evolution, it must also include evidence for transitional stages in the process of descent with modification from older forms to newer forms in a given ancestor-descendant lineage of organisms.

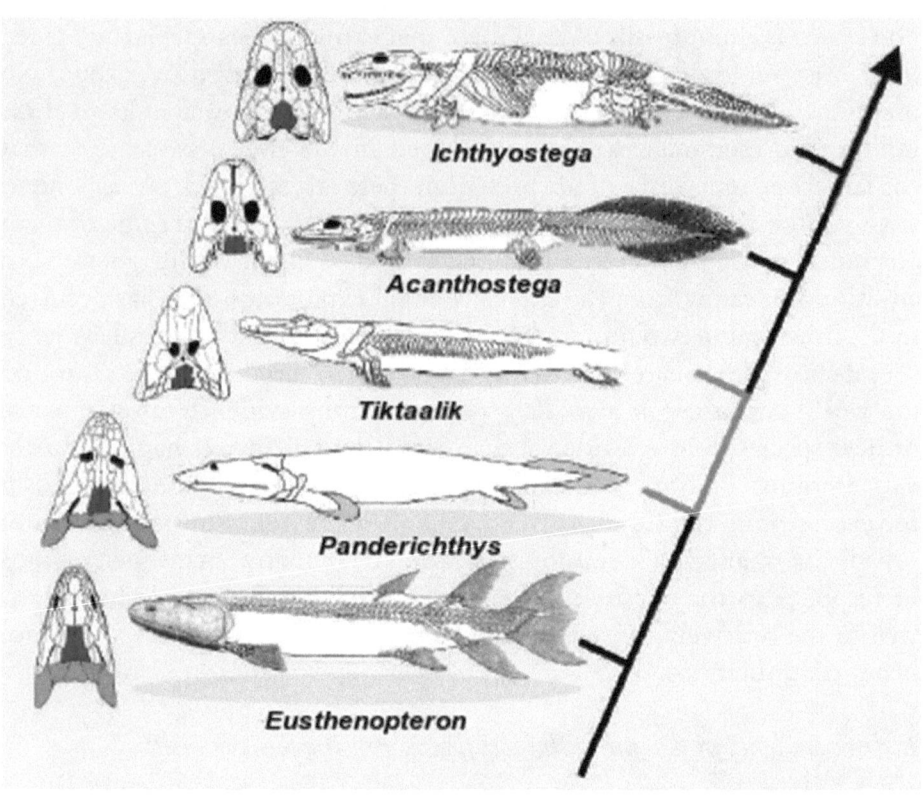

Figure 4.2. *Tiktaalik* in context. Five of the most completely known fossil transitional animals between fishes and land animals are shown. *Eusthenopteron* was a lobe-finned fish. *Panderichthys* and *Tiktaalik* are transitional forms. *Acanthostega* and *Ichthyostega* were land animals. Note the loss of a gill cover (light gray) in the skull roof of *Tiktaalik*. Reprinted with permission from Macmillan Publishers Ltd. *Nature* 440: 747 (copyright 2006)

In April 2006, the scientific journal *Nature* reported a remarkably complete fossil of an extinct creature—part fish, part land animal—that was named *Tiktaalik rosea* (figure 4.2). The find, which had been made in 2004, was exciting to evolutionary biologists for two major reasons. The first of these is that *Tiktaalik* is clearly a missing link between extinct transitional creatures that were more fishlike in appearance and functioning and those that were more amphibian-like. *Tiktaalik* has fishlike features not present in the amphibian-like fossils, like fin rays at the ends of the front legs, and amphibian-like features not present in the fishlike fossils, like front-leg bone structure and rib structure that allowed it to push itself up and drag along in shallow water. The second reason that the *Tiktaalik* find was exciting was because of the age of the rock in which it was found. The more fishlike fossils had been found in rock about 385 million years old while the more recent amphibian-like fossils were dug out of rock less than 377 million years old. Besides being intermediate in form between these fossils, the *Tiktaalik* fossils were discovered in rock *intermediate in age* between them (Ahlberg and Clack 2006; Daeschler *et al.* 2006). In fact, the scientific team that made the discovery did their geological research before beginning their trek to find remote dig sites, away from structures that might impede their work and, more importantly, where the rock was between 377 and 385 million years old. The radiometric dating of rock done by others previously did not fail them, nor did the theory of descent with modification. They found transitional fossils in rock laid down right smack in the time interval where they should have been in order to fit the evolutionary sequence. *Tikaalik rosea* lived about 382 million years ago (Ahlberg and Clack 2006; Monoyios 2008).

If you visit the Carnegie Museum of Natural History in Pittsburg, Pennsylvania, USA, you will find a wonderful display of horse evolution that shows the complete record, beginning with the small five-toed *Hyracotherium* that lived between 50 and 60 million years ago right up to *Equus domesticus*, the modern horse, with all known intermediates between. *Hyracotherium* had teeth suited for munching twigs while modern horses have teeth suited for grass. The fossils found in rock spanning the ages between these two species show a transition from twig-munching teeth to grass-chomping ones. Transitional fossils are all from rock that places them all in the correct sequence with respect to degree of similarity to/difference from *Hyracotherium* and *E. domesticus*. You'll also see the transitional hoof forms. The fossil record is equally complete as a cornucopia of evidence for descent with modification of elephants from ancestral protoelephants (Shoshani 1997, 38; Handwerk 2008), whales from ancestral hippopotamus (hoofed land animals) (Futuyma 2005, 78–79), mammals from ancestral reptiles (Futuyma 2005, 76–78), birds from theropod dinosaurs (Futuyma 2005, 73–76), modern humans from apelike ancestors (Futuyma 2005, 79–83), and many other lineages of living organisms.

## 2. Morphological evidence for common ancestry in living creatures: mammalian limbs

*Do mammals share a unique common ancestor? What do morphology and the fossil record say?*

A mammal is an animal with a segmented backbone and hair (never feathers) made from the protein keratin. Human beings are mammals. Do all mammals share unique common ancestry? That is, do they all descend from a far-distant common ancestral population of animals that were the first animals with a backbone and hair? If they do, then some unique features that all mammals share in common and that only occur in mammals are evidence that all mammals share a unique common ancestral population. In other words, all mammals and only mammals share a unique common ancestor.

If mammals do not share unique common ancestry, then there are two possibilities. One is that there are other animals, *but without* backbones and hair made of keratin, *that also descended* from an ancestor that had a backbone and hair made of keratin. Another possibility is that mammals evolved *more than once*. So we have three possible scenarios as diagramed in figure 4.3: (a) all mammals and only mammals share a unique common ancestor, (b) there are other animals without backbones and without hair made of keratin that also descended from the same ancestor from which mammals descended, and (c) the condition of having both a backbone and hair made of keratin evolved more than once.

JOSEPH FORTIER

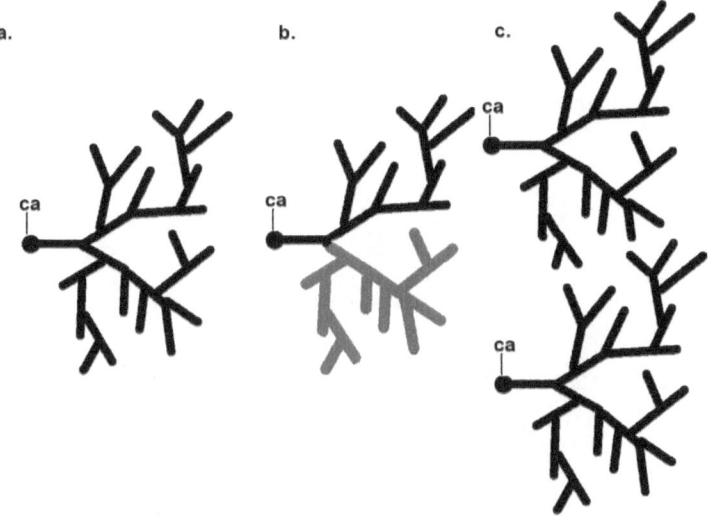

Figure 4.3. Three possible scenarios for mammal evolution. Black lines represent mammalian evolutionary lineage, with branching evolution as the lineage diversifies into various species. Gray lines represent a hypothetical non-mammalian lineage evolving from the same common ancestor as that of mammals. (A) All mammals and only mammals share a unique common ancestor. (B) There are other animals without backbones and without hair made of keratin also descended from the same ancestor from which mammals descended. (C) Mammals, with the unique condition of having both a backbone and hair made of keratin, evolved more than once. (Note: ca = common ancestor.)

Now let's take a look at the facts to see which of the three scenarios in figure 3 is best supported by evidence. Let's start with scenario A: all mammals and only mammals share a unique common ancestor. Figure 4.4 shows that the forelimb bones of four very diverse kinds of mammals have the same anatomical pattern. Whether it is a human being with forelimbs adapted for grabbing, holding, and lifting (a); a whale with forelimbs adapted for swimming (b); a horse with forelimbs adapted for running (c); or a bat with forelimbs adapted for flying (d), the bones show the same pattern. In all, there is a humerus (h), radius (r), ulna (u), carpals (m), and phalanges (p). This similar pattern of bone structure, in spite of a wide variety of functioning across these groups of mammals, is found in all mammals, even when a mammal's use of the front limb is more birdlike than say, doglike or fishlike. Thus, bats' front limb bone structure is more doglike than birdlike. The bones in whales' front flippers are more doglike than fishlike. This is evidence that supports our hypothesis that all mammals, in spite of their diversity, are descended from a common ancestor. Their various bones have remained as those in the ancestor, but each has become shaped differently as a result of natural selection over large amounts of time to meet the peculiar functioning that has emerged in each present-day mammal group.

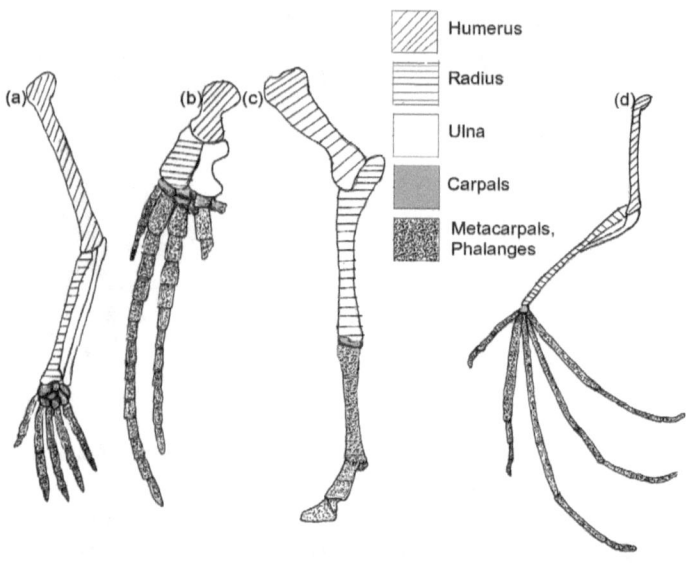

**Figure 4.4. Forelimbs of a human, whale, horse, and bat. Even though the functions of the forelimb on these organisms are widely different, the same forelimb bones remain present in these organisms—evidence for their shared common ancestry.**

Here's a clincher: the name *mammal* comes from the unique mammary glands that the females of all mammals have for providing milk for the young. Whether we're dealing with a platypus mother that lays eggs; a kangaroo or a marsupial mouse that carries the fetus in a pouch after a certain stage of development; or a human, a whale, or a bat that carries the fetus in the uterus, we're dealing with animals in which the mother provides her young with milk from mammary glands. No other creature in the world has mammary glands other than mammals, just as no other creature in the world has the combination of hair, a backbone, and limbs as we've seen. All this is evidence for a unique common ancestry.

Now, let's look at scenario B in figure 4.3. To date, no type of creature has been discovered that has the combination of backbone, hair made of keratin, forelimb bone structure described above, but no female mammary glands—in other words, not as described above. There is no evidence for this scenario. If we ever find some, it will cause scientists to redefine what

JOSEPH FORTIER

it means to be a mammal. The same is true for the scenario represented by C in figure 4.3. There is no evidence that any of these structures evolved twice, in different groups of animals, but as we shall see, there is lots of evidence that supports just one evolutionary mammalian lineage. In fact, with bone structure, the situation is the reverse. Birds, other reptiles, and amphibians all have forelimb bone structure closely similar to that of mammals. These creatures, in turn, have forelimb bone structure more similar to that of *Ichthyostega* (figure 4.2) than to that of other Devonian vertebrates in the aquatic-to-land fossil transition. Thus, the fossil evidence suggests that mammals arose from a group of reptiles from the Permian period (Futuyma 2005, 76–78) (figure 4.1), which in turn arose from amphibians that had arisen from earlier transitional terrestrial creatures from the Devonian period such as *Ichthyostega* (Ahlberg and Clack 2006; Futuyma 2005, 73).

*Vestigial structures*

Although modern whales have front limbs that reveal their former history as land mammals, they have no well-defined rear limbs. However, the fossil record shows that their whale ancestors did have nonfunctional but clearly defined remnants of a pelvis disconnected from the backbone and remnants of leg bones connected to the vestigial pelvis bones to boot. Two such fossil whale and porpoise ancestors were *Dorudon*, which lived 41–33 million years ago, and *Basilosaurus*, which lived 40–34 million years ago (Futuyma 2005, 79; Campbell et al. 2007, 251; Sutera 2000; Bergen Museum, *Whale Pelvis* and *Whale Evolution*).

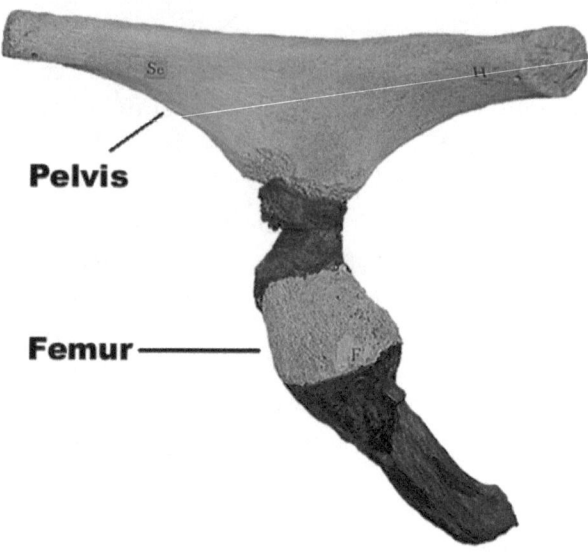

**Figure 4.5. Atavistic pelvis and femur of a Northern Right Whale. Although modern whales have no external hind limbs, fossil evidence shows that their ancestors did. The reappearance in some modern whales, in reduced form, of structures that occurred in fossil ancestors is additional evidence that whales evolved from land animals that used their hind limbs for walking on land. Courtesy of Bergen Museum, University of Bergen/Terje, Lislevand.**

Extinct ancestors of modern animals are not the only creatures with vestigial remnants of hind pelvises and legs (Darwin 1859, 490–496). Some modern whales, in fact, have atavisms of those ancestral hind pelvic limbs. An *atavism* is a reappearance of a characteristic that was present in a remote ancestor but has not been evident in intervening ancestors. Examples and photos of these atavisms in various modern whales can be found at the Bergen Museum website (Bergen Museum, *Whale Pelvis*), such as the one in figure 4.5. Among living snakes, the boas and pythons have vestigial hind pelvic and leg remnants (Francisco 2001, 41). These vestigial and atavistic structures in whales neither had nor have a locomotory function. In boas and their relatives, they have derived a secondary function as claspers in sexual copulation. Whale atavistic legs are relics of whale ancestors that walked around on land. The same is true of boas and pythons—they evolved from four-legged ancestors as other snakes did (Norris 2007, *NewScientist* 2006). Their vestigial hind legs are the feet sticking out from under the bed, which gave the story away, along with other evidences such as the Jacobson's organ, a smell organ in the mouth that snakes share with lizards. In 2007, a fossil lizard was reported with vestigial *front* legs but perfectly well developed hind legs. The 95-million-year-old fossil belongs in the lizard group most closely related to snakes (Norris 2007).

Humans also have a number of vestigial structures, including the coccyx at the end of the spine (Fig. 4.6). The coccyx is a vestigial tailbone.

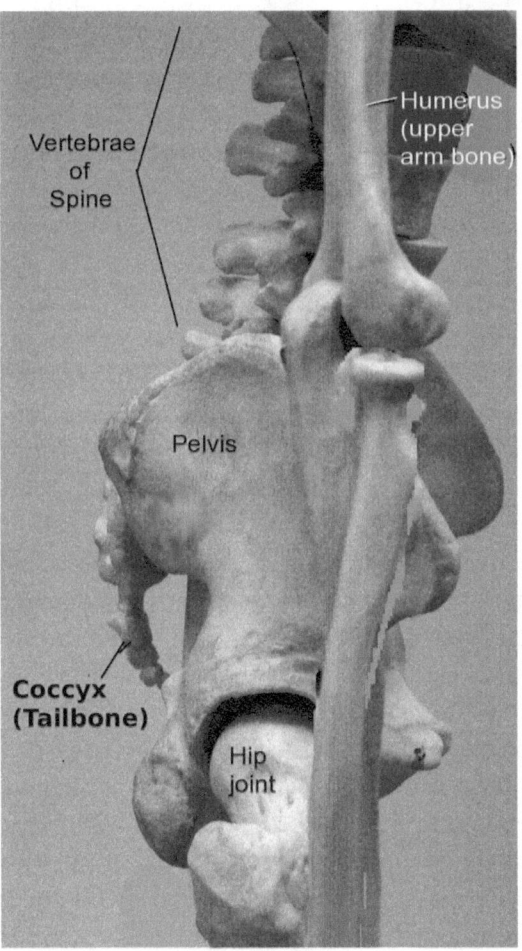

**Figure 4.6. Diagram of the human pelvic area, right side. The coccyx is a vestigial tailbone at the lower end of the spinal column.**

# 3. Biochemical evidence

*Human insulin and bacteria*

Thanks to Watson, Crick, and others, we now have the ability to read the language of DNA as we saw in chapter 3. We can also read the protein language spelled out by amino acid sequences as we also saw in chapter 3.

After the structure of DNA was discovered, applied scientists called genetic engineers started working on techniques to build DNA molecules so that the language encoded in these DNA molecules would say what they wanted it to say. In 1978, a team of genetic engineers working for a biotechnology company called Genentech began building artificial genes that would code for human insulin. Insulin, synthesized by the pancreas, is a protein that regulates blood-sugar levels. Pancreas damage or dysfunction results in a deficiency of insulin production, which leads to type 1 diabetes, an incurable disease in which sufferers must dose themselves with insulin for the rest of their lives. The goal of the work was to develop an artificial insulin-coding gene that could be inserted into the chromosomes of *E. coli* bacteria, a common bacterial type that occurs in the human intestine and is easy to rear in the lab. Why insert a human gene into an *E. coli* bacteria? After all, these are completely different creatures. Could one creature possibly translate the DNA language into proteins in exactly the same way as another? Wouldn't the result of such an insertion be, at best, a significantly different order of amino acids in the protein sequence of the bacterial product? As it turned out, in 1982, when *Humulin* was patented and put on the market as the first recombinant DNA drug approved by the US Food and Drug Administration, *E. coli* bacteria—huge vats of them—had demonstrated that by the billions, they all translate the human DNA sequence into amino acid sequences of proteins *exactly* as human cells do. The insulin churned out en masse by the *E. coli* vats did not differ in the least from human-produced insulin (Campbell et al. 2007, 222). Humans and bacteria—and for that matter, all living things—have the same DNA, differing only in sequence among species. The function of the DNA is exactly the same in all species of life with a few very minor exceptions among a handful of groups of microorganisms, and the way that the DNA base sequence is read is exactly the same in all species. It seems that early on, life stumbled onto something that worked pretty well, and as it diversified into various forms during eons of time by natural selection and descent with modification, there was never much of a reason to change the way that the helping proteins read the DNA code and transcribed it into RNA. Certainly, were evolution not the explanation for the uniform reading of the genetic code of all life, the odds would be infinitesimally tiny that this uniformity would be the case.

*Evo-devo and the case for unique common ancestry of all animals*

As we saw in chapter 3, Morgan and Sturtevant, in their lab in New York at the turn of the twentieth century, developed various mutations in the fruit flies they worked with. The geneticist's love of the fruit fly has by no means diminished, even in spite of all the tricks these geneticists can do with DNA. A growing branch in evolutionary studies over the past few years is evo-devo, or evolutionary developmental studies. Evo-devo studies the evolution of genes involved in the developmental processes of various organisms.

One of those genes that "wakes up" during the pupal (cocoon-like) stage of the fruit fly is called the "eyeless" gene. Modern evo-devo geneticists have retained Morgan and Sturtevant's custom of referring to fruit fly genes by the effects of the mutant allele. The normal allele of the *eyeless* gene produces a protein that starts a cascade of activity among perhaps hundreds of genes that code for various aspects of adult fly eye development. Without the embryonic *eyeless* gene, these other genes would never become activated.

*A word on the fly life cycle.* The life cycle begins when the mother fly lays eggs. Out of each egg hatches a tiny larva—in the case of a fly, a maggot. The maggot eats and grows until it is ready to make a pupa, a cocoon-like structure. In the pupa, the maggot tissues break down, a group of embryonic genes is activated, and the creature inside the pupa undergoes an eerie metamorphosis into a fly. Each embryonic gene awakens a cascade of genes, which together order the production of completely new tissues and organs. The product: an adult fly.

In 1994, a team of evo-devo practitioners found that in a fly pupa, the embryonic gene that activates the gene cascade responsible for building the fly's eyes is amazingly similar to a gene (*aniridia*) in the human embryo. It turns out that the human embryonic gene also activates a cascade of genes that builds human eyes. Furthermore, they found that an equally amazingly similar gene in the mouse embryo not only turns on mouse eye development, but also turns on *fly eye development*. The gene that produces these closely similar developmental proteins is now known as the

PAX-6 gene. Amino acid sequence segments of fly and human PAX-6 gene protein products are shown in figure 4.5. In fact, the human PAX-6 gene is nearly identical to that of the mouse. The PAX-6 gene has been found in animals as simple as flatfish, where it is associated with the development of very simple eyespots, and in mollusks such as squid. It has been found in sea squirts, a sedentary intertidal animal in which the immature or larval form has a notochord as in vertebrate embryos, and in fish (Quiring 1994; Gehring 1997, 2004; Zuker 1994; Strickler 2001; Carroll 2005, 66–69). An explanation is that this gene originated in the common ancestor of all animals at least at the level of complexity of the flatworm. The complexity of a flatworm is just a step above that of jellyfish—not very complex. It follows, using the logic of descent with modification, that it has been conserved in the descendants of that first ancestral animal population.

| GENE | SEGMENT OF PROTEIN (AMINO ACID SEQUENCE) |
|---|---|
| eyeless (fruit fly) | THYPDVFARERLAAKIDLPEARIQVWFSNRRAKWRREE |
| aniridia (human) | THYPDVFARERLAGKIGLPEARIQVWFSNRRAKWRREE |

Figure 4.7. Developmental PAX-6 genes of a fruit fly (top) and human being (bottom), and segments of each respective gene's protein products. Letters symbolize various protein amino acids. Underlined gray letters symbolize amino acids that differ in the two amino acid sequences. Developmental genes named at left produce proteins (segments shown) that initiate cascades of other genes, which in turn are responsible for eye development.

The alternative would be, as Sean Carroll put it in his book *Endless Forms Most Beautiful*, that this gene has been called upon to build eyes from scratch in a wide array of animals (Carroll 2005). A remarkable coincidence, to say the least. The problem, of course, is that the laws of chance preclude this alternative. Gene lengths of animals, even as simple as a flatworm, are measured by the number of kilobase pairs in the gene's DNA. One kilobase equals one thousand base pairs. The odds of two independent emergences of a gene, both of which have identical or nearly identical base-pair sequences, is small to the point of nonexistence. The oldest bacterial fossils are about 3.5 billion years old. The oldest evidence for animal life is the 635-million-year-old chemical remains of ancient sponges found in Oman, southeast of Saudi Arabia (Love et al. 2009). A fossil flatworm from Alaska was found in rock between 600 and 680 million years old (Allison 1975).

JOSEPH FORTIER

Presuming that this flatworm represents the earliest date in which the PAX-6 gene might be present, the shortest amount of time that it took for the PAX-6 gene to evolve would be from the age of rock, in which those first bacterial fossils were found, to the age of the Alaskan fossil, or from 3.5 billion years to 680 million years (for the most conservative estimate). This represents a time span of 2.82 billion years. It took life a minimum of 2.82 billion years to come up with the first PAX-6 gene. All other major groups of animal life arose during the Cambrian period, between 530 and 505 million years ago. In order for a gene that took 2.82 billion years or 2,820 million years to evolve once, it would have had to evolve again in only 25 million years, which is 0.008, or eight thousandths, of the time it took for the first appearance. To be built from scratch three or four, perhaps five or more times within eight thousandths of the time it took for the PAX-6 gene to evolve once? Not likely.

The close similarity of the PAX-6 gene in animals with three cell layers in their embryos—such as flatworms, squid, fruit flies, sea squirts, fish, mice, and human beings—is evidence that these animal groups descended from a common ancestor.

The improbability is compounded by the fact that there are other embryonic genes, such as *Distal-less* (*Dll*), that behaves as PAX-6 does. Distal-less is named, in true Morgan-Sturtevant fashion, for the effect of its mutant allele on the fruit fly. Fruit flies homologous for the *Dll* recessive (with both alleles at the locus recessive/mutant) don't have the ends of their appendages. Sean Carroll and his team found that the dominant, normal allele turns on limb development in not only fruit flies but also in other arthropods like butterflies, crustaceans (crabs, lobsters), spiders, and centipedes. Pretty exciting, not to mention fun. Next he tried placing the gene in embryonic cells of chickens, fish, sea squirts, marine worms, and sea urchins. He got animal parts, which were appropriate enough to the respective animals, but growing from wildly inappropriate body regions of the animals, sticking out where they didn't belong. Chicken feet developed where chickens don't have feet, fish fins where fish don't have fins, siphons on sea squirts where they don't belong, "legs" (parapodia) of sea worms in the wrong places, and tube feet on sea urchins in inappropriate places. Amusing in a twisted sort of way, but how could this be explained? These animals, although occurring in widely different places on the animal evolutionary tree, all have an embryonic gene closely similar to *Dll*, just as *eyeless* is closely similar to *aniridia*—the same gene for all practical purposes. In all these animals, the fly version (*Dll*) was close enough to their embryonic genes that turn on limb development that *Dll* expressed itself in them, just as their own version would if mischievously misplaced in the embryo (Carroll 2005).

So now it seems we have two genes that both had to first evolve once over about 2.73 billion years and then both evolve again four or five times from scratch within 25 million years.

Since for each of these genes, the second evolution from scratch would have to occur in 8/1000 of the time it took for the first evolution, this means that the chance of both of them simultaneously doing this would be $(8/1000)(8/1000) = (8/1000)^2$, or .000064. This number is 64/1,000,000 (sixty-four millionths). It is a small number. A small probability.

The evidence strongly suggests that the reason these genes are so similar in widely divergent groups of animals, animals whose base sequences in their other DNA is also widely divergent (as one would expect), is because these particular developmental genes have been strongly conserved during evolutionary history.

*Comparison of the human and chimpanzee genomes*

The Human Genome Project was begun in 1990 and completed in 2003. It was coordinated by the US Department of Energy and the National Institutes of Health. Participants from twenty-four other nations contributed to the project, including Great Britain, Japan, France, Germany, and China. The project estimated the number of human genes to be between 20,000 and 25,000. A maximum estimate of the number of these genes that code for proteins is 24,500. Other subsequent studies estimated this number of human genes to be between 42,000 and 70,000. Altogether, there are 3 billion base pairs in the DNA of the human genome (Martin 2008).

The chimpanzee genome was published in an article in 2005 by the Chimpanzee Sequencing and Analysis Consortium, a scientific team supported by the National Human Genome Research Institute. The article was published in *Nature* in 2005. The study found that the genomes differed by about 150 million base pair differences throughout the respective genomes. Not only that; while chimps have twenty-four chromosome pairs, humans have only twenty-three. The study found that chimps, like humans, have about 3 billion base pairs throughout their entire genome, that is, all the DNA from all the genes on all the chromosomes (Li 2005; The Chimpanzee Sequencing and Analysis Consortium 2005). Now let's put these numbers in context. First, to find the percentage difference in base pairs between chimp and human genomes, let's start by setting up a proportion:

a.

$$\frac{150,000,000}{3,000,000,000} = \frac{x}{100}$$

b.

$$x = \frac{15,000,000,000}{3,000,000,000}$$

c.

$$x = 5\%$$

We want to find what percentage (x) 150 million is of 3 billion. By multiplying both sides of equation (a) by one hundred, we arrive at the value of $x$ (b and c), and find that the percentage of base pairs that are different between the chimp and human genomes is just 5 percent. They are 95 percent similar. So in spite of these huge numbers of DNA base pairs we're dealing with in whole genomes, these two genomes are in fact closely similar.

*Humans, chimpanzees, and the number of chromosome pairs*

A key difference between the great apes (chimpanzees, orangutans, gorillas) and humans is the number of chromosomes. While the great apes all have twenty-four chromosomes, humans have only twenty-three. Is this disparity evidence against shared common ancestry of humans with great apes? It certainly seems so at first glance. After all, chromosome number is a very basic defining characteristic that separates groups of living organisms.

The area of biology known as *karyology* investigates the structure of chromosomes in various organisms. In 1991, research conducted by a team of scientists at the Yale School of Medicine led by Dr. J. W. Ijdo was published (Ijdo et al. 1991). The study produced some interesting and telling results that provide insight into the answer to the above question.

Before we look at their study, we need to first look at some basic characteristics of chromosomes. When chromosomes are viewed under a microscope, lateral bands can be seen on them. These bands vary in thickness, distance from each other, and color (light/dark) so that each kind of chromosome can be recognized by its distinct pattern of banding. Each kind of chromosome can also be recognized by its shape. Cytologists (scientists who study cells) and geneticists use these characteristics as clues to identify each kind of chromosome within a cell. For example, humans have twenty-three pairs of chromosomes. Each chromosome in a pair has a banding pattern and shape that exactly matches that of the other chromosome in the pair. Thus, scientists refer to human chromosomes as chromosome pair one, chromosome pair two, and so on up to pair twenty-three.

Another characteristic of chromosomes, whether human, plant, or bacterial, is that they have tail or telomere regions at one end of each chromosome. Telomeres can be recognized by their regularly repeating sequences of base pairs. They function to protect the chromosome from becoming mutated during replication.

In the study conducted by Dr. Ijdo and his colleagues, they found that each chromosome in human chromosome pair two consists of what used to be two separate chromosomes. At the point where these two chromosomes were fused, there is a remnant of two separate but overlapping chromosome segments, each containing remnants of what used to be each chromosome's telomere. Furthermore, each half of human chromosome two has a banding pattern and shape that matches a specific chimpanzee chromosome (chimp chromosomes twelve and thirteen) (Sun et al. 1978 a, b). Thus, the evidence shows that somewhere along the human evolutionary pathway from the common ancestral population shared by chimpanzees and humans, two chromosomes fused in the human lineage at their telomere ends, forming one chromosome (chromosome two).

## 4. Embryonic evidence

An embryo is the stage of development between first cell division and birth, hatching, or germination, depending on the kind of living organism. Actually, we have already seen some strong embryonic evidence for common ancestry among animals when we looked at animal evo-devo or common genetic sequences and functions of those sequences among diverse animal lineages.

In 1828, before Darwin published *Origin of Species*, the eminent developmental biologist Karl Ernst von Baer took notice that embryonic features of larger, more generalized categories of living organisms, such as "all vertebrates (animals with a backbone and dorsal nerve cord)" appear earlier in embryonic development than specific features unique to smaller, specialized categories (e.g., "all vertebrates that live on land only"). This generalization is still recognized as von Baer's law. For example, embryos of all vertebrate land animals, including humans, display pharyngeal slits, a notochord, body segmentation, tails, and paddle-like limb buds before features more characteristic of their more specific category become apparent. In land animals like snakes, chickens, and people, the pharyngeal slits don't develop into gills as they do in fish (Futuyma 2005). In humans and other mammals (vertebrates with hair at some stage and in which females nurse their young with milk from mammary glands), they develop into the middle ear canal and various glands in the neck region (Phillips 1991, 291–296).

The notochord, along with the dorsal nerve cord, is found in active forms of primitive chordates (the larger category of animals to which vertebrates belong), such as larval sea squirts (tunicates) and adult cephalochordates, all of which are aquatic. In these organisms, the notochord, composed of cartilage, is unsegmented, lies dorsally in the body, and has the dual function of a locomotory organ and a support for the dorsal nerve cord. The notochord maintains body rigidity and shape and provides leverage around which the segmented muscles can work to provide propulsion through the water (Waggoner 1995).

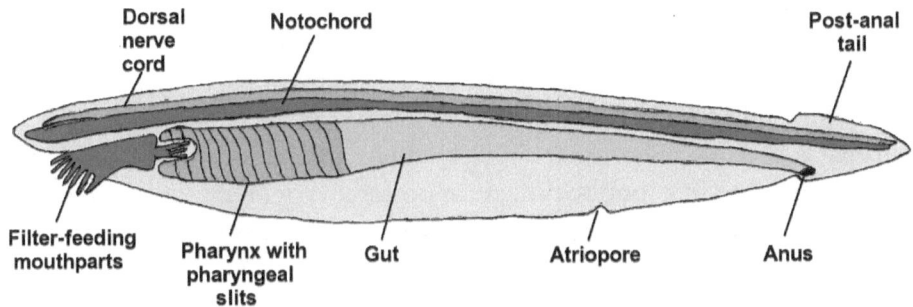

**Figure 4.8. Diagram of anatomy of a cephalochordate, a primitive member of phylum Chordata, to which all vertebrates, or animals with a vertebral column, also belong. Note the four cardinal characters of all chordates, including vertebrates, at some stage of their life cycles: notochord, dorsal nerve cord, pharyngeal slits, and post-anal tail.**

Cephalochordate adults (figure 4.8) have a pharynx, or filter-feeding device, perforated with many pharyngeal slits. Water passes in through the animal's mouth. In the pharynx/pharyngeal slit assembly, microorganisms and other small food particles are filtered out of the water into the digestive tract, and the water passes out through the pharyngeal slits, and eventually out of the body through a ventral atriopore. Behind the anus lies a tail with segmented musculature (Waggoner 1996).

Interestingly, and in accord with von Baer's law, all these major features appear at least in embryonic form in higher vertebrates. As we've seen above, the pharyngeal slits appear in embryonic form in all vertebrates although they differentiate into gills in fish and various glands and other organs in land animals. In most vertebrates, except for very primitive, jawless ones like the lamprey, the notochord degenerates after functioning in the embryo. In these embryos, it induces development of the central nervous system before degenerating.

Fish, like primitive chordates, have an extensively segmented muscular system, as any fish eater will readily note. In land animals, muscular segmentation is reduced or lost. However, retention of the theme of segmentation within the vertebrate body plan is readily observed in the vertebral column.

Embryos of all vertebrates, including tailless ones such as great apes and humans, have postanal tails, homologous to that of cephalochordates.

A signature of the group Chordata (including vertebrates) is the dorsal nerve cord. In all other animals without either notochord at some stage in their lives or vertebral column, in those in which a nerve cord exists, it is always ventral. Examples are insects, lobsters, flatworms, and earthworms.

Thus, in chordates, including vertebrates, certain structures unique to chordates such as pharyngeal slits, postanal tails, notochords, and dorsal nerve cords that may not be present in postembryonic life stages are found universally in embryonic stages and provide evidence for unique common ancestry of all chordates, including humans.

## 5. Speciation and evidence for its occurrence

The process of speciation is fundamental to biological evolution, or descent with modification. All evidence for evolution is implicit evidence for speciation. Without speciation, there is no evolution as reflected in the fossil record and as understood as the engine of diversification of life. What is speciation? It is the process by which new forms of life or species, modified by environmental pressures on heritable traits of organisms in a population, descend from these older, ancestral species populations. "Descent with modification" means evolution or emergence of new forms of life, which are modified from preexisting forms. Is there evidence that this really happened and is happening and is not just abstract theory?

Below, we'll investigate the theory of how new species come into existence and take a look at evidence that supports this theory.

*Allopatric speciation: the theory*

Allopatric speciation occurs when genetic barriers arise between populations of a species. It is considered to be by far the most common way that new species evolve. Populations may become physically separated from each other by the development of geographical barriers such as mountain formation and rivers or a habitat barrier, such as a desert. Since barriers such as these prevent the members of one segment of these separated populations from mating with members from the other segment, gene flow by intermating between the populations is effectively stopped. Over generations, genetic differences between the two populations multiply until even if members of the populations do come into contact again, their genetic differences will be too great to mate and produce healthy, fertile offspring at a rate sufficient to homogenize the two populations again into one. Still, if these genetic barriers that arise as a consequence of the geographical barriers are not sufficiently strong when and if the two populations again come into contact, a hybrid zone may develop in which limited mating with limited success occurs between pairs composed of members from each population (Futuyma 2005, 381).

What is a genetic barrier? How would certain kinds of genetic differences decrease the likelihood of healthy, fertile offspring? To answer this question, we need to review what alleles are. Remember that an allele is a version of a given gene. An allele for fruit fly eye color codes for red eye color while another allele at that same locus (site) on the pair of homologous chromosomes codes for white eyes. Now, say there are two different populations of a kind (species) of animal (could also be a kind of plant, fungus, or bacterium). Say at a given locus on a specific chromosome, one population of this animal (let's make it a chipmunk) is homozygous for allele $A_1$. That is, in all individuals, each of the two homologous chromosomes (one maternal and the other paternal) on which this gene occurs has only the $A_1$ version, or allele, of the gene—never $A_2$. Thus, for the homologous chromosome pair, the site (locus) on each chromosome at which the gene occurs will have only the $A_1$ allele for every individual chipmunk in the population. Thus, each chipmunk will have two $A_1$ alleles: $A_1A_1$. The other chipmunk population is likewise homozygous but for allele $A_2$. On the homologous chromosome pair on which this gene occurs, the gene version (allele) is always $A_2$. Each chipmunk in this population will have two $A_2$ alleles (one at each locus for that gene on each homologous chromosome on which that gene occurs): $A_2A_2$. In each of these populations, one of the alleles of the gene ($A_1$ or $A_2$) has become fixed: $A_1$ in population one and $A_2$ in population two. There is only one allele of that gene in the gene pool of each population. Of course, such genetic fixation normally takes a lot of time and a lot of generations.

Now as it turns out, let's say chipmunks with the heterozygous condition ($A_1A_2$, rather than $A_1A_1$ or $A_2A_2$) have low reproductive success. In chipmunk population one, either because allele $A_1$ has some slight advantage over $A_2$ or because it just so happened that when the original ancestral population was split and diverged into populations one and two, each resulting population was quite small, and there were quite a few more $A_1$ alleles in population one as a whole than $A_2$ alleles. Perhaps there was a combination of both of these effects. Ditto for chipmunk population two. The low fitness of the heterozygote condition $A_1A_2$ is in fact a genetic barrier to these two populations becoming one again. That is, after several generations, if a female chipmunk from population one with $A_1A_1$ genotype meets the chipmunk of her dreams from population two with $A_2A_2$ genotype and they mate, all their offspring will have the heterozygous $A_1A_2$ condition and will either die early or not have many or any offspring if they survive long enough to mate. Thus, the two chipmunk populations are somewhere along the road to becoming completely genetically isolated and distinct species.

Now consider the possibility (probability, given a long enough time span in which the populations are isolated geographically) that other genetic loci in these two populations also become different, such that the heterozygote condition at these loci also results in low reproductive success. There may well be a multiplier effect, further causing the chance of reproductive success to be low to nonexistent from matings between populations.

*Allopatric speciation: evidence*

a. *Appalachian salamanders*

Salamanders are little animals shaped like lizards, with little legs and a long tail, but more closely related to frogs. Like frogs and unlike lizards, they have to lay their eggs in the water and prefer to stay where it's moist.

They come in a lot of sizes and colors. Hellbender salamanders can get over a foot long, while the world's smallest salamander is only about an inch long. Chinese giant salamanders get over three feet (one meter) long.

Stephen Tilley and a team of fellow scientists looked at the effects of sexual isolation on reproduction among Allegheny Mountain dusky salamanders in the southern Appalachian Mountains of the eastern United States (they're only a few inches long). These mountains probably have the highest salamander species diversity in the world. It isn't uncommon to find a salamander species that only exists in one remote valley in these mountains. Tilley and the team caught dusky salamanders from various isolated populations among these mountains, kept track of which individuals were members of which population, and set up experiments to test and compare mating success between individuals from the same versus different populations. They kept score of the proportion from the same versus different populations that when brought together mated. What they found was that among the various pairs of populations, the effects of geographical isolation on sexual attraction varied continuously with distance between populations, from almost 100 percent mating between pairs to almost no mating between pairs. Furthermore, when they compared genetic similarity by looking at DNA, they found that consistently, the more geographically distant the populations, the more genetically different they were. Thus, the increasingly smaller likelihood of mating of pairs taken from increasingly more distant populations was correlated with increasingly less genetic similarity (Futuyma 2005, 381; Tilley et al. 1990). The implication is that the modicum of genetic similarity necessary for sexual attractiveness between populations of these dusky salamanders varies directly with geographical distance between the populations. Because of the physical barriers to mating between populations that the mountains impose, the greater the distance between populations, the slimmer the likelihood of genetic mixing (by mating) between those populations. The slimmer this likelihood, the greater the likelihood that enough genetic differences had built over time so that both mating attractiveness and likelihood of healthy offspring between individuals from respective distant areas became increasingly minimal.

b. *Devastating and migratory grasshoppers*

The devastating grasshopper, *Melanoplus devastator*, lives in the western coastal states of the United States west of the Sierra Nevada and Cascade mountain ranges. These grasshoppers feed on a number of kinds of plants and are destructive to rangeland forage, orchards, grains, vegetable crops, and gardens (Pfadt et al. 1994a).

The migratory grasshopper, *Melanoplus sanguinipes*, is widespread throughout the western United States east of the Sierra Nevada and Cascade mountain ranges. While it tends to be less generalist than the devastating grasshopper, concentrating its feeding on grasses and forbs including grain crops and rangeland crops, it will feed on orchard and other crop plants besides grains. It is considered the most destructive North American grasshopper to agriculture (Pfadt et al. 1994b).

Matthew Orr from the University of California at Davis and a team of other scientists found a valley in the California Sierra Nevada east of Sacramento where the two grasshopper species above mate with each other. This came as a surprise since the two species are quite easy to tell apart and the male sex organs are quite different. In fact, it is not only obvious from a morphological standpoint that they are different species, but it is also considered important from an agricultural standpoint to maintain their distinctness since their feeding habits are quite different. Yet both molecular evidence and morphological evidence from individuals of each species, and their hybrid progeny caught in the mountain pass, show that these species intermingle and mate in this one small area in a pass in the California mountains. Furthermore, Orr and colleagues crossed devastator and migratory grasshoppers in the lab. They caught these grasshoppers far from the mountain pass where their ranges met. The progeny of these lab-reared and mated grasshoppers were closely similar to hybrids caught in the mountain pass.

Although the two species can interbreed where their ranges meet, the egg mortality of these crosses was 50 percent greater than that of eggs resulting from matings between individuals of either species. Furthermore, in the lab, grasshoppers mated much more readily with mates of their own species than with the other species.

Thus, a likely scenario is that that the two populations from which the two species evolved were geographically separated by the mountain range for a long enough period that genetic barriers to reproduction developed. But when individuals from each species came in contact again at the mountain pass, they were able to mate and reproduce—although not very efficiently. This fact presents itself as evidence that there was, at one time, one ancestral population from which the two species descended.

*Incipient speciation in South American fruit flies*

As long ago as 1962, it was shown that the fruit fly *D. paulistorum* was actually a complex of six different populations on the road to becoming new species. Crosses between three of these populations from Central America, the Amazon, and the Andes of southern Brazil resulted in sterile male hybrids. The cause of the male sterility was the presence of a foreign chromosome in the hybrid mother, which altered the egg structure, causing sterility in males that developed from the eggs. Thus, male sterility was observed to act as a fertility barrier between these three populations, a mechanism that would eventually result in three species (Ermann 1962).

Speciation is a process that occurs over long periods of time—longer than a person's lifetime, even for creatures with short generation times like fruit flies. However, the cause-and-effect relationships that have been proposed in theory between accumulation of genetic differences and sterility barriers between populations that at one time were not different on the one hand and the evidence we've seen that supports this theory on the other together comprise another piece of the puzzle that snugly fits with the other pieces we have seen, all of which support descent with modification from common ancestry.

# Summary

To answer the question, what is the evidence that evolution has happened and is happening? we looked at five categories of evidence: (1) fossil evidence, including radiometric dating of rock in which fossils are found, (2) morphological evidence, (3) biochemical evidence, (4) embryonic evidence, and (5) ecological and molecular evidence for presently ongoing speciation.

☐ From fossil and radiometric evidence, we found that all over the face of the earth, in any given layer of sedimentary rock with similar age, the same kinds of fossils are found in only that layer. Successively deeper layers of rock in the earth's crust correspond with increasingly older rock and with fossils increasingly different from creatures we see today. We also saw that there is abundant evidence in the fossil record for transitional forms of life, such as *Tiktaalik*, a transitional form between fishlike and amphibian-like fossils. The fossil evidence is that the evolution of life occurred over vast amounts of time and that descent with modification is supported.

☐ From morphological evidence, we found that structures such as limb bones, mammary glands, and hair all support descent with modification from common ancestry of diverse aquatic, flying, and walking creatures such as whales, bats, horses, and humans. Furthermore, we found that front limb bone structure supports descent with modification from common ancestry of the above creatures with birds, other reptiles, amphibians, and transitional creatures such as *Tiktaalik*. In turn, evidence supports descent with modification of *Tiktaalik* and all terrestrial vertebrates and vertebrates descended from terrestrial vertebrates from common ancestry with fish.

☐ We also saw the evidence for the common ancestry of whales and land animals and of snakes and four-legged land animals provided by vestigial structures, such as the vestigial pelvic fins in fossil and modern whales and vestigial legs in boa constrictors and pythons.

☐ From biochemical evidence, we've seen that all life, including simple creatures such as *E. coli* bacteria and humans, use the same language coded in DNA. Humans have put this powerful, persuasive piece of evidence for the common ancestry shared by all life to practical application by "cutting and pasting" the human gene for insulin production into the chromosome of *E. coli* bacteria and harnessing huge colonies of the bacteria to manufacture human insulin.

☐ The new field of evolutionary development (evo-devo) has uncovered exciting new evidence for the remote common ancestry of all animal life. Embryonic genes that turn on complex cascades of other genes that instruct the developing embryo in building complex structures such as eyes have been recently discovered. One of these embryonic genes, the PAX-6 gene, called the eyeless gene in fruit flies and the aniridia gene in humans, is so closely similar in mice, humans, and fruit flies that when the mouse PAX-6 gene is cut and pasted into a fruit fly chromosome, it functions perfectly in triggering fruit fly eye development. The PAX-6 gene has also been found in other animals such as squid, flatworms, and sea squirts.

☐ The findings from the human genome project and chimpanzee genome project show that about 95 percent of human DNA is identical to chimp DNA. It was thought that a key genetic distinction between the chimp and human genome was that while chimps and other tailless nonhuman primates have twenty-four chromosomes, humans have only twenty-three. It is now known that human chromosome number two is a fusion of what in chimps remain chromosomes twelve and thirteen. Half of human chromosome two is identical to chimp chromosome twelve and the other half to chimp chromosome thirteen.

☐ Besides the molecular evidence provided by evo-devo for the common ancestry of animal life, other embryological evidence that we saw supports the unique common ancestry of all animals that have a notochord at some point in their development. At some point in their development, all members of this animal group, which includes humans, uniquely possess the following: a notochord, a postanal tail, pharyngeal slits, and dorsal nerve cord.

☐ Finally, we've looked at some of the evidence for presently ongoing speciation: the splitting of a species into two (or more) new species. We've seen examples in which genetic reproductive barriers are presently in the process of being developed between populations of closely similar animals (populations of dusky salamanders, populations of grasshoppers, and populations of fruit flies), in which these populations have been geographically separated for long periods of time.

☐ The evidence we have looked at is diverse: everything from rock layers to radioactive measuring devices to shapes of organisms to their embryos to their genes to their ecological settings. The consonance of each piece of evidence with all the evidence taken together reliably and consistently supports Darwin's theory of new species emerging by descent with modification from unique, common ancestry over long periods of time as heritable traits are acted on by environmental factors over generations.

# Glossary

**allopatric speciation**. The process by which new species emerge from an ancestral population because of the isolation, over many generations, of two or more subpopulations of the ancestral population, in which the isolation is caused by a geographic barrier.

**amber**. Fossilized plant resin or sap.

**atavism**. The reappearance of a characteristic in a modern or more recent extinct organism that was present in a remote extinct ancestor but was lost in intervening extinct ancestral forms.

***Basilosaurus***. An extinct ancestor of modern whales and porpoises that lived 40–34 million years ago.

**carbonization**. The process by which some fossils of plant and insect remains are fossilized, in which the fossil is composed of the black carbon remains of the organism.

**cephalochordate**. Small eel-like animals related to vertebrates that live in sandy areas near water and filter small organisms from water on which they feed.

**Chordata**. The large, general category of animals in which at some life stage the following four structures are present: gills slits, notochord, postanal tail, and dorsal nerve cord (lying close to the back of the animal). Vertebrates are the largest subcategory of the Chordata.

**chromosome**. A DNA molecule and the assortment of various molecules that assist it in its various functions, such as replication and gene expression. The chromosome consists of functional segments called genes.

**DNA**. The large macromolecule in cells of living organisms that is the information database that governs the growth, development, form, and function of the organism and is capable of replication. Because of the DNA molecule's capacity for replication, its information is passed on to future generations or organisms.

***Dorudon***. An extinct ancestor of modern whales and porpoises that lived 41–33 million years ago.

**element**. A kind of substance that is only composed of one kind of atom.

**embryo**. An early developmental stage of an organism.

**evo-devo**. The biological discipline of evolutionary development studies in which the evolution of genes involved in the developmental processes of various organisms is studied.

**fossil**. A rock impression of a formerly living organism, piece of an organism, or other evidence for its former existence.

**gene**. A functional segment of a chromosome consisting of a segment of the chromosome's DNA and associated molecules that assist in the gene's expression.

**genetic barrier**. The emergence of an inability to mate and produce fertile offspring between males and females of what was formerly a species because of the accumulation of genetic differences among two or more populations of that former species.

**genome**. The sum total of all genes in an organism.

**geological timescale**. A chart of the ages of rock strata often including names of fossils characteristic of each stratum.

**half-life**. The amount of time that it takes for half the amount of a radioactive element to decay into the stable isotope that is its decay product.

**Humulin**. The human gene for insulin production that was genetically engineered into the genome of the bacterium *E. coli* so that insulin can now be mass-produced by large cultures of *E. coli*.

***Ichthyostega***. An extinct species of primitive tetrapod considered to be one of the first land vertebrates.

JOSEPH FORTIER

**igneous**. A type of rock that is formed when molten lava cools and solidifies.

**isotope**. A variety of a given type of substance called an element.

**karyology**. The subdiscipline within biology that investigates the structure of chromosomes in various organisms.

**keratin**. The type of protein of which hair and the fingernails or claws of mammals consist.

**lead 207**. A stable form of the element lead; also the decay product of uranium 235.

**maggot**. The wormlike immature life stage, or larva, of certain groups of insects, such as flies and wasps.

**metamorphic**. A type of rock that is formed from igneous and/or sedimentary rock when it is crushed under high temperature and pressure.

**notochord**. A rigid structure composed of cartilage that appears in the early embryonic development of vertebrates and is retained in various life stages of primitive vertebrates and their relatives.

**orogeny**. The scientific study of mountain formation.

**PAX-6 gene**. A gene in the embryos of many major groups of animals, including humans and fruit flies, that initiates the process of eye development.

**permineralization**. A process of fossilization that occurs when a living organism is buried in sediment, usually at the bottom of a body of still water. Minerals dissolve out of the sediment particles, replacing dead organic matter in the organism's remains.

**pharyngeal slits**. Narrow indentations that appear behind the heads of vertebrate embryos in their early development.

**pupa.** The quiescent life stage between the larval stage and the adult stage in certain groups of insects in which there is little or no movement and larval organs and body form are reorganized into those of the adult form.

**radiometric dating.** A method used for dating the age of rock in which the amount of an unstable radioactive substance found in the rock is compared with the amount of a stable isotope that is a known decay product of that radioactive substance also in the rock.

**sedimentary.** A type of rock that was formed when sediments such as sand, silt, or clay particles settled to the bottom of a body of still water, forming layers of rock over long periods of time.

**shale.** A type of fine-grained sedimentary rock formed when the predominant sediment type that settles to the bottom of still water is clay.

**speciation.** The process of emergence and radiation of distinct species from ancestral populations over evolutionary time.

**telomere.** A segment at one end of the chromosome that protects it from being mutated.

**_Tiktaalik rosea._** An extinct species of transitional creature that is a missing link between extinct fishlike creatures and extinct amphibian-like creatures. The first _Tiktaalik rosea_ fossils were found in the Canadian Arctic in 2004.

**uranium 235.** The radioactive form of the heavy element uranium.

**vertebrate.** Animals with a backbone, including humans.

**von Baer's law.** The general observation that embryonic features of large, general categories of organisms, such as vertebrates, flatworms, and flowering plants, appear earlier in embryonic development than specific features. These more specific features are unique to smaller subcategories (such as apes) nested within the larger categories (vertebrates).

# References

Ahlberg P. E. and J. A. Clack. 2006. "A Firm Step from Water to Land." *Nature* 440: 747-749.

Allison C. W. 1975. "Primitive Fossil Flatworm from Alaska: New evidence Bearing on Ancestry of the Metazoa." *Geology* 3: 649-652

Campbell N. A, J. B. Reece, and E. J. Simon. 2007. *Essential Biology Third Edition*. San Francisco, CA, USA: Pearson Benjamin Cummings.

Carroll S. 2005. *Endless Forms Most Beautiful*. New York: W. W. Norton & Co.

Canadian Fossil Discovery Centre. 2009. "Fossilization." Last modified September 28, 2011. http://www.discoverfossils.com/education/fossilization.html

The Chimpanzee Sequencing and Analysis Consortium. 2005. "Initial Sequence of the Chimpanzee Genome and Comparison with the Human Genome." *Nature* 437: 69-87

Daeschler E. B., N. H. Shubin, and F. A. Jenkins. 2006. "A Devonian Tetrapod-Like Fish and the Evolution of the Tetrapod b=Body Plan." *Nature* 440: 757-763

Dalrymple G. B. 1991. *The Age of the Earth*. Palo Alto, CA, USA: Stanford University Press.

Donovan S. K. 1991. *The Processes of Fossilization*. New York: Columbia University Press.

Francisco L. R. 2001. "Chapter 5: Class Reptilia, Order Squamata (Ophidia): Snakes." In *Biology, Medicine, and Surgery of South American Wild Animals*, edited by M. E. Fowler. and Z. S. Cubas. Ames, Iowa. Iowa State University Press.

Futuyma D. J. 2005. *Evolution*. Sunderland, MA, USA: Sinauer Associates, Inc.

Gehring W. J. 1998. *Master control genes in development and evolution*. New Haven, Connecticut. Yale University Press.

Gehring W. J. 2004. Master control genes in development and evolution. In *Biennial Report 2002-2003*: Biozentrum. http://www.biozentrum.unibas.ch/report0203/index.html

Grimaldi D. M. and M. S. Engel. 2005. *Evolution of the Insects*, New York: Cambridge University Press. h2g2. 2003. "Whale Evolution—the Fossil Evidence." In *A Hitchhiker's Guide to the Galaxy*, British Broadcasting

Company. Last modified 2010. http://www.bbc.co.uk/dna/h2g2/A1283186.

Handwerk B. 2008. "Ancient Elephant Ancestor Lived in Water, Study Finds." Last modified October 2010. http://news.nationalgeographic.com/news/2008/04/080414-elephant-evolution.html.

Ijdo J. W., A. Baldwin, D. C. Ward, S. T. Reeders, and R. A. Wells. 1991. "Origin of human chromosome 2: An ancestral telomere-telomere fusion." *Proceedings of the National Academy of Science* 88: 9051-5

Li W.H. and M.A.Saunders. 2005. "The chimpanzee and us." *Nature* 437: 50-1

Lincoln R. J, G. A. Boxshall, and P. F. Clark. 1993. *A Dictionary of Ecology, Evolution, and Systematics*. New York: Cambridge University Press.

Love G. D., E. Grosjean, C. Stalvies, D. A. Fike, J. P. Grotzinger, A. S. Bradley, A. E. Kelly, M. Bhatia, W. Meredith, C. E. Snape, S. A. Bowring, D. J. Condon, and R. E. Summons. 2009. "Fossil steroids record the appearance of Demospongiae during the Cryogenian period." *Nature* 457: 718-21

MacRae A. 2007. "Burgess Shale Fossils." Last modified 2007. http://www.geo.ucalgary.ca/~macrae/Burgess_Shale/

Martin S.A. 2008. "Human Genome Project Information." U.S. Department of Energy Office of Science, Office of Biological and Environmental Research, Human Genome Program. Last modified July 25, 2011. http://www.ornl.gov/sci/techresources/Human_Genome/home.shtml

Matthews W. H. 2005. *Roadside Geology of Southern British Columbia*. Missoula, Montana USA: Mountain Press Publishing Company. 24.

Monoyios K. 2008. The Search for Tiktaalik. Chicago: University of Chicago http://tiktaalik.uchicago.edu/searching4Tik.html

Bergen Museum. 2009. "Whale Pelvis." Bergen, Norway: University of Bergen. Last modified December 17, 2009. http://bergenmuseum.uib.no/fagsider/osteologi/hvaler/e_bekken.htm

Bergen Museum. 2009. "Whale Evolution." Bergen, Norway: University of Bergen. Last modified December 17, 2009. http://bergenmuseum.uib.no/fagsider/osteologi/hvaler/e_evolusjon.htm

Kimm Groshong. 2006. "Oldest Snake Fossil Shows a Bit of Leg." In *New Scientist*. Last modified April 19, 2006. http://www.newscientist.com/article/dn9020-oldest-snake-fossil-shows-a-bit-of-leg.html

Norris S. 2007. "Weird Lizard Fossil Reveals Clues to Snake Evolution, Experts Say." In *National Geographic News*: http://news.nationalgeographic.com/news/2007/03/070326-lizard-snakes.html

Orr M. R., A. H. Porter, T. A. Mousseau, and H. Dingle. 1994. "Molecular and Morphological Evidence for Hybridization between Two Ecologically Distinct Grasshoppers (*Melanoplus sanguinipes* and *M. devastator*) in California." *Heredity* 72: 42-54

Pfadt R. 1994. *Field Guide to Common Western Grasshoppers*. Last modified October 28, 2010. http://www.uwyo.edu/grasshoppersupport/html_pages/fieldgde.htm

Phillips J. B. 1975. *Development of Vertebrate Anatomy*. Saint Louis, Missouri: The C. V. Mosby Company.

Quiring R. 1994. "Homology of the Eyeless Gene of Drosophila to the Small Eye Gene in Mice and Aniridia in Humans. *Science* 265: 785-789

Shoshani J. 1997. "What can make a four-ton animal a most sensitive beast?" *Natural History* 106: 36-45

Strickler AG. 2001. "Early and late changes in Pax6 expression accompany eye degeneration during cavefish development." *Developmental genes and evolution* 211: 138-144

Sun, N. C, C. R. Sun, and T. Ho. 1978a. "Chimpanzee chromosome 12 is homologous to human chromosome arm 2q." *Cytogenetic and Genome Research* 22: 594-597

Sun N. C., C. R. Sun, and T. Ho. 1978b. "Chimpanzee chromosome 13 is homologous to human chromosome arm 2p." *Cytogenetic and Genome Research* 22: 598-601

Sutera R. 2000. "The Origin of Whales and the Power of Independent Evidence." *The Reports of the National Center for Science Education* 20: 33-41.

Tilley S. G., P. A. Verrell, and S. J. Arnold. 1990. "Correspondence between sexual isolation and allozyme differentiation: A test in the salamander *Desmognathus ochrophaeus*." *Proceedings of the National Academy of Science USA* 37: 2715-9.

Waggoner B. 1995. "Chordata: More on Morphology." Last updated October 4, 2011. http://www.ucmp.berkeley.edu/chordata/chordatamm.html

Waggoner B. 1996. "William Smith (1769-1839)" Last updated October 4, 2011. http://www.ucmp.berkeley.edu/history/smith.html.

Waggoner B. 1996. "Introduction to the Cephalochordata." Last updated October 4, 2011. http://www.ucmp.berkeley.edu/chordata/cephalo.html.

Zuker C. S. 1994. "On the evolution of eyes: would you like it simple or compound?" *Science* 265: 742-743.

# CHAPTER 5

## What Positions Do People Take with Respect to Evolution and Religious Faith?

IN CHAPTER 1, we saw that the Genesis creation narratives, when taken literally in their cultural context, do not contradict evolution. Just as other ancient Near Eastern people had no conception of the scientific method or of evolution, neither did the authors of these narratives. They assumed a structure of the universe in common with all ancient Near Eastern people and incorporated this cosmology into their creation myths. These creation myths, as we saw, were intended to be neither scientific nor historical. They convey a deeper, timeless understanding of human nature, the nature of God, and his relationships with the universe and with people. This timeless understanding remains unique and valuable to people of faith all over the world up to the present age. For example, the number seven held literary symbolic significance to ancient Near Eastern people. It represented completeness, fulfillment. The ancient non-Hebrew Mesopotamian writers of the *Enuma elish* wrote their document on seven clay tablets. The first biblical creation narrative occurs over seven days. In both narratives, the final day is a day off, signifying completion, fulfillment. The first biblical creation narrative makes use of the first six days as a rhetorical device to repeat major theological themes: "God said let there be . . . and so it was" and "And God saw that it was good." God is portrayed as using the Jewish lifeway of a six-day workweek and a seventh day of rest to convey that the Jewish way of life is good and Godlike. To the Hebrew people, their way of life was about their relationship with God.

We also saw that the second biblical creation narrative in Genesis is concerned with timeless, ahistorical issues such as the relationships between innocence, sexuality, and wisdom (culture, the world, self-promotion) and how this complicated set of relationships affects the human relationship with God and freedom. We saw in this story that God (*Yahweh*) is portrayed as compassionate, humble, and also humanlike. The name of the first man, Adam, is the Hebrew word for "earthman." In the story, it is associated with the Hebrew word for *soil* or *earth*, which is closely similar to the word for *man*. Thus, the author did not intend to give the impression that he was referring to a historical personage; he used the earthman as a character in his story to convey truths about us all from time immemorial. Thus, the Adam and Eve narrative is another creation myth intended not to convey history but rather to convey timeless, existential truths about relationships between God and humans. Again, the writer (about four hundred years before the final writer of the first creation narrative) was not attempting to write human paleontology. Thus, there is no conflict between this narrative and biology. The ancestral population of humans had to be sufficiently large to assure sufficient genetic diversity so that the following generations could be genetically healthy. If there were not enough genetic diversity in enough individuals at the outset of the human species, then over the first several generations, inbreeding would lead to frequent expression of unhealthy genes, genetic diseases, and extinction.

It would seem that the case would be closed and that there would be nothing more to say about any conflict between evolution and faith. This is not the case. Does this mean that people who believe in God tend to be less reasonable or less educated than people who do not? Does it mean that believers tend to be less intellectually honest? Or is it the other way around? Are those who accept evolution and interpret evolutionary theory as pointing to a completely pervasive randomness, chance, and lack of meaning in the universe shortsighted? Do they, like the proverbial ostrich putting its head in the sand, avoid necessary nonscientific questions that people ask and wonder over?

Before these issues can be addressed, it is first necessary to consider the types of positions that people take regarding evolution and faith. In his book *When Science Meets Religion*, Ian Barbour sorts the variety of ways in which people relate science and religion into four major categories: conflict, independence, dialogue, and integration (Barbour 1999). I will discuss types of positions taken regarding evolution and faith using Barbour's fourfold typology.

## Conflict

With respect to evolution, two major positions fall within the conflict category. These are the following: (1) a biblical literalist approach sometimes used today in which the reader assumes a contemporary meaning to the words of scripture and (2) scientific materialism, or scientism. Those who ascribe to the former position do not take into account the findings of modern studies of ancient Near Eastern languages and archaeology, such as those we summarized in chapter 1. For purposes of brevity, we'll refer to this position as ahistorical biblical literalism. Those who ascribe to the second position assert that the most fundamental reality in the universe is matter. They assume that the only way to know anything is by the scientific method, which investigates the behavior of matter. As we saw in chapter 2, empiricism is a theory of knowledge that asserts that knowledge arises from sensory perception of the evidence concerning matter. Scientific materialists deny the credibility of any knowledge other than empirical knowledge. It's important to make clear that many or most scientists are not scientific materialists. While scientists use the tools and intellectual methods of scientific empiricism when they are investigating scientific phenomena, most would also agree that there are areas of knowledge not accessible by these tools and methods. Because scientific materialists deny the existence of such knowledge, they believe that questions that cannot be addressed with the scientific method are not answerable or, as often put, irrelevant. Scientific materialists are by definition atheists or agnostics since God transcends accessibility by the methods of scientific empiricism. Where adherents to both of these positions do agree is in their mutual belief that the science of evolution and religious faith are not compatible and can never reach agreement. Both cannot be true, in their opinion. Let us first look at ahistorical biblical literalism and then move our focus to scientific materialism.

*What are the concerns and positions taken by ahistorical biblical literalists?*

From the point of view of religious people who adopt this position, there are three major issues at stake: (a) the challenge to biblical literalism, (b) the challenge to human dignity, and (c) the challenge to design.

a. *Biblical literalism: historical background*

In the fourth century AD, the Christian scholar Augustine of Hippo stated that when a conflict arises between biblical scripture and demonstrated knowledge, scripture should be interpreted metaphorically. About the creation narratives in the book of Genesis, he said that scripture is not concerned with the "form and shape of the heavens," and "does not intend to teach [people] things not relevant to their salvation" (Augustine, 42–43; Barbour, 8). Galileo Galilei, a preeminent astronomer and physicist who in the seventeenth century elaborated on the theory that the earth circles the sun rather than vice versa, was summoned by the Catholic Church's

Inquisition to explain his heresy. The church's doctrine at the time, based on its (ahistorical) interpretation of various passages of the Bible, held that the earth is at the center of the universe and all celestial bodies orbit the earth. Galileo defended himself with Augustine's quote (above) and by asserting that we can learn from two sources: the "Book of Nature" and the "Book of Sacred Scripture." Both, he said, are from God, so they can't contradict each other. However, the Inquisition convicted him as a heresy suspect, sentenced him to prison (later commuted to house arrest), and banned his publications. In the twentieth century, Pope John Paul II expressed regret for the Galileo affair and how Galileo was treated and officially conceded that Earth was not stationary (*NewScientist* 1992). In March 2008, the Vatican proposed to complete its rehabilitation of Galileo by erecting a statue of him inside the Vatican walls (Owen 2008). In December of the same year, during an event to mark the four hundredth anniversary of Galileo's earliest telescopic observations, Pope Benedict XVI praised his contributions to astronomy (BBC 2008).

At the time Darwin's *Origin of Species* was published, theologians who rejected all forms of evolution were in the minority. Most theological conservatives accepted evolution, although reluctantly, accepting metaphorical rather than ahistorical literal interpretations of the Genesis narratives. Other theologians enthusiastically endorsed evolution, claiming it as God's way of creating and noting its consistency with their rosy view of historical progress. These people were open and enthusiastic about the emerging field of biblical scholarship that included studies of ancient Near Eastern languages and literature (Barbour 1999).

Christian Fundamentalism emerged in the United States at Princeton Theological Seminary in the first decade of the twentieth century. It spread among Protestant denominations during and just after the First World War. The purpose of the movement was to zealously defend Protestant Christianity against liberal theology and Darwinism (Noll 1992). The first formulation of American fundamentalist beliefs can be traced to the Niagara Bible Conference (1878–1897). During these years, a fourteen-point creed was established. The first of these points states the fundamentalist position on biblical inerrancy, including that "we understand . . . that the Holy Ghost gave *the very words* [italics mine] of the sacred writings to holy men of old" (Sandeen 1967; Unger 1981, 320). These fourteen points, in whole or in part, may be found verbatim in the doctrinal statements of various contemporary evangelical congregations (Founders Ministries 2011; St. Germain 2009; Covenant Community 1997).

After 1910, the Fundamentalist movement became more sharply defined. Historian George Marsden described it as "militantly anti-modernist Protestant evangelicalism." High among its priorities was the "inerrancy of Scripture," which, given the movement's historical development, evidently meant literal interpretation of its words unaided by modern biblical scholarship. World War I intensified the Fundamentalists' sense of impending crisis in Western civilization. The term *fundamentalism* was coined in 1920 by Baptist editor Curtis Lee Laws to refer to those ready "to do battle for the Fundamentals" (Noll 1992).

Catholicism was not immune from fear of evolutionary theory. In the 1947 papal encyclical *Humani Generis*, Pope Pius XII stated, "Some imprudently and indiscreetly hold that evolution, which has not been fully proved even in the domain of natural sciences, explains the origin of all things, and audaciously support the monistic and pantheistic opinion that the world is in continual evolution" (Pope Pius XII 1947, no. 5). The same fear of loss of awareness of the personal God as haunted the Fundamentalist Christian denominations was expressed in the document. Further along, the pope stated that Catholics may not believe that humans have descended from an original ancestral population but rather from two ancestors, one named Adam (Pope Pius XXI 1947, no. 37). However, in the same document, the pope stated that "the teaching authority of the Church does not forbid that . . . research . . . take place with regard to the doctrine of evolution, in as far as it inquires into the origin of the human body as coming from pre-existent and living matter" (Pope Pius XII 1947, no. 36). The Jesuit priest-scientist Pierre Teilhard de Chardin was silenced by his Jesuit superior general and by the pope for his writings on evolution and Christian theology in which he integrated the two areas. Only after he died in 1955 were Teilhard's written works allowed by Rome to be published in Catholic-sponsored presses and to be taken off the forbidden book list. By 1996, Pope John Paul II affirmed evolution, acknowledging its veracity and, paraphrasing Galileo, stated that "truth cannot contradict truth" in his address to the Pontifical Academy of Sciences on October 1996 (Pope John Paul II 1996).

b. *Evolution and human dignity*

A deeply held precept of monotheistic faiths in a personal god is that human persons are uniquely created in the image and likeness of God. This religious truth is taken from the Genesis first creation narrative in which human creation occurs on the sixth day. The implication by Darwinist thought that humans have evolved in a natural process by random, undirected accident flies in the face of this belief in intentional human creation, one on which human dignity hangs in these traditions. In such a directionless world without human distinctness from other creatures and without a purposeful, relational God, theists fear the loss of the sense of human dignity on which morality, purpose, and meaning in human life are predicated. It is easy to understand the difficulty for many Jews, Christians, Muslims, and others of faith that this seeming conflict presents. Unless Darwinian science is sufficiently addressed by the theology of a personal god, it would be hard to see how monotheistic faiths in a personal, relational god who affirms human dignity could survive the scientific evidence. This vital issue will be addressed.

## c. *Evolution and purposeful design*

In a worldview without change, an intelligent designer seemed persuasively argued for in the complex ways in which organisms can function and in their diverse adaptations in form and function to harmoniously live in their respective environments. Once Darwin and those who followed him explained how adaptation could be accounted for by an impersonal, random process of variation and natural selection, the universe suddenly seemed devoid of purpose (Barbour 1999, 9–10).

Seyyed Hossein Nasr, a distinguished Muslim scientist and religious thinker, delivered the prestigious Gifford Lectures in 1981 on the topic "Knowledge and the Sacred." In the lecture, he maintained that before the development of modern mechanistic science including Darwinism, the sense of the unquestioned sacred principle of being that caused and causes existence allowed people a meaningful cosmos. Nature was perceived as being endowed with an order and meaning that vividly expresses its divine origin. Since the place of the divine source has been replaced by many modern thinkers with the profane causal past, these thinkers find it difficult to perceive nature any longer as the visible manifestation of eternal, absolute reality and meaning (Haught 2000, 65–66).

A consequence of this perceived loss of the sacred origins of nature that Nasr sees is the massive environmental destruction we are currently witnessing, including life systems. The desacralizing of nature implied in the loss of a sense of sacred origins has made it difficult or impossible for the modern mentality to any longer perceive inherent value in life. As a result of this devaluing of life, our planet is left vulnerable to economic greed and reckless exploitation. So Nasr claims that our current ecological crisis is the inevitable result of modern science's elimination of a sense of sacred origins and diminishing of the value of life (Haught 2000, 66).

*What do scientific materialist positions look like?*

As we saw above, materialism is a philosophical position that asserts that matter is the fundamental reality in the universe. Empiricism is a position that asserts that knowledge is acquired by applying the rational mind to our sensory experience of matter. Scientific materialists deny the credibility of any knowledge other than empirical knowledge. Another aspect of most forms of materialism is reductionism. Reductionists assert that the laws and theories of all sciences, including biology, psychology, and sociology, are reducible to chemical and physical laws. Furthermore, the component parts of any system (such as living organisms) determine its behavior (Barbour 1999, 11–12).

Among scientists who adopt scientific materialism as their sole way of viewing the world, the tendency is to diminish the value of religion as a way to knowledge since religion is perceived as lacking the objective, reproducible data that science produces. Unlike the scientific enterprise, religious thinkers don't use experimental testing and other means of empirical measurement and are thus subjective, close minded, and uncritical in this view (Barbour 1999, 11–12). I will describe the conflict positions of two preeminent scientists, the first of whom takes a reductionist position and the second of whom does not.

Edward O. Wilson, a preeminent entomologist (insect specialist), champion of biological diversity conservation, and Pulitzer Prize–winning author, provides an example of the thinking of scientific materialists. In his book *Sociobiology*, he asks how altruistic behavior could arise in social insects such as ants if such behavior compromised the reproductive future of the individual. He traced the genetic origins and evolutionary progression of social behavior in ants, other animals, and humans. Wilson demonstrates how seemingly altruistic behavior enhances the survival of the individual's close relatives and, thus, maximizes the probability that the genes of these relatives, many or most shared by the altruistic individual, will be successfully passed on. From these observations, Wilson leaps to the conclusion that all human behavior can be reduced to and explained by biological origins and genes. He states that "it may not be too much to say that sociology and the other social sciences, as well as the humanities, are the last branches of biology" to be subsumed under the umbrella of evolution and genetics as the ultimate explanation. Eventually, he continues, the mind will be explained as "an epiphenomenon of the neural machinery of the brain" (Barbour 1999, 13; Wilson 1975, 4).

Even religion will be seen to be a biological epiphenomenon according to Wilson. In his book *On Human Nature*, Wilson holds that religion is a sort of vestigial human trait, harkening back to a time in human natural prehistory when religious practices were useful survival mechanisms, contributing to group cohesion. He states that the power of religion will be gone forever when religion is explained as a product of evolution. It will be replaced by a "philosophy of scientific materialism" (Barbour 1999, 13; Wilson 1978, chap. 7 and 8).

It is interesting that E. O. Wilson grew up in a Christian Fundamentalist home. According to his autobiography, *Naturalist*, he was an ardent young Southern Baptist during his adolescence, teaching Sunday school for his church. Later, he felt he had to make a choice between science and religion and chose science, which he felt required that he abandon his faith (Wilson 1995).

Stuart Kauffman, a member of the Santa Fe Institute (www.santafe.edu) during the 1990s, presently on the faculty of the University of Calgary, and MacArthur Fellowship (Genius Award) recipient, provides a fascinating updated example of a conflict position that rejects some key aspects of traditional scientific reductionism and the Darwinian notion that complex organization emerged purely by chance and contingency. We'll take a more focused look at his scientific thinking in the next chapter. As we'll see in chapter 7, the Jesuit anthropologist and geologist Pierre Teilhard de Chardin, in the first half of the twentieth century, anticipated Kauffman's empirical findings on emergence of complex structures by about fifty years. While Teilhard finds transcendent spiritual meaning in the large pattern of emergences of complex entities with their own unique, nonreducible laws that emerge with them, Kauffman seems to studiously avoid such meaning.

Kauffman provides some of his rationale for rejecting religious faith in his book *Reinventing the Sacred*. He explains that post-Enlightenment secular thinkers have been scandalized so that they feel that the very words *sacred* and *God* are "utterly corrupted." There are two major sources of this scandal: the sanctioning of murder ("death") "in the name of God" and the "aggrandizing certainty of religious fundamentalists." There is a fear by these thinkers that the "sacred" may involve becoming totalitarian. Kauffman wishes for a "global spiritual space" in which honest, open dialogue between faiths and civilizations concerning the meaning of sacred can take place (Kauffman 2008, 282). He seems unaware that such dialogue is already in operation among Christian denominations and between Christianity and Asian religions such as Hinduism and Buddhism. Nonetheless, what he suggests is vitally important among a wide array of faith and religious traditions and among as many of the world's civilizations as possible and, of course, also including secular input. A major objective of this book is to encourage such dialogue, using biological evolution as a pretext.

Kauffman is of course correct that religious institutions have been responsible for atrocities perpetrated against those of other faiths and cultures and that this awful tendency continues to this day. However, he fails to distinguish between the human nature of the perpetrators of such evils within these institutions and the content of the religious faith these people and institutions supposedly represent. Religious institutions and their representatives carry with them considerable influence and potential power. This influence and power has been, is, and will be abused as long as people and institutions are what they are, whether they be religious or secular. Also, he fails to acknowledge the wealth of significant contributions that religion has made to human existence, including in the areas of art, science, social work, health, and education.

I remember a history instructor of mine, Fr. John O'Malley, SJ, who said that the history of the Catholic Church is a history of constant reform. Of course, there would be no need for reform if there were not mistakes from which to reform. The same might be said of the many modern governments including that of the United States, which saw its way to reform its racist legal structures once it was properly instructed by its African American religious leadership, especially in the person of Rev. Martin Luther King Jr. Kauffman defines human free will as a sort of Aristotelian "unmoved mover" (Kauffman 2008, 198). Perhaps it is possible to use free will to choose to conflate imperfect, flawed moral human behavior with the ideals of religious faith on the one hand or on the other to make this important distinction and see flawed moral behavior as abuse of religious faith.

In his first book on the science of complexity, Kauffman expressed compassion and respect for the moral anguish and sense of loss of the sacred that underlies "creation science" (Kauffman 1995, 6). In his second book, he claims that those he calls "Christian fundamentalists," with their "aggrandizing certainty," are a cause of the scandal that modern post-Enlightenment secularists experience (Kauffman 2008, 282). I wonder why the transition from the compassionate understanding expressed in the 1995 book to the harsh judgment expressed in the writing of the 2008 book. Also, I wonder why a first-rate scholar like Kauffman would not have done his homework in mainstream Christian theology concerning science and religion, such as Douglas Futuyma did in his book *Science on Trial* (Futuyma 1995, 19–20). Jay Gould himself recognized that Catholic thinking is receptive to biological evolution (Gould 1997). Stuart Kauffman, while admitting to the same point with respect to Pope John Paul II's 1996 letter to the Pontifical Institute of Sciences concerning biological evolution, confuses *soul* with *mind* in his reference to that document. The pope wrote, "If the human body takes its origin from pre-existent living matter, the spiritual soul is immediately created by God" (John Paul II, 1996). It is confusing why Kauffman substituted *mind* for *soul* (Kauffman 2008, 177).

Kauffman shows himself capable of a dialogue position. He describes the position of "some Jesuit astronomers" who see God as "less omniscient and omnipotent," which he claims is remarkably close to what he is discussing. But he continues, seemingly illogically, to claim from this that "even if this God exists but cannot know, this God cannot reliably answer prayer" (Kauffman 2008, 283); thus, he remains consistent with his conflict position. In fact, these same Jesuits hold that God and God's power is transcendent mystery that even the tremendous power of the human mind and the power of language cannot ever completely capture. They hold that evidence shows that God's relationship with the universe is not that of a dictator but rather that of a wise, loving creator-leader, with respect for the growing process and need for freedom—and care—of the creation's process of becoming. This position will be discussed further in chapters 8 and 9.

Kauffman seems to assume the Greek all-in-control god in Barbour's sense (Barbour 2000, 150–153), whom he describes (not exclusive to the Greek notion) as the supernatural generator god who is the source of all the vastness around us. Kauffman's notion of God is, in contrast, pantheistic: "nature itself." He asks, "What more do we really need of a God?" We are, Kauffman says, finally responsible, to the best of our forever-limited wisdom (Kauffman 2008, 283). In subsequent chapters, these issues will be addressed.

# Independence

Independence positions aim to avoid conflict. The strategy, conscious or unconscious, used by independence positions for conflict avoidance between science and religion is to avoid any semblance of dialogue or interaction between these two areas of human thought and experience. Barbour notes that conflict is avoided, which is a good starting point for the larger project of finding the sort of unity and integrity perhaps aimed at by John Paul II (truth cannot contradict truth). The methods, questions, and functions of religion and science are distinct, and these distinctions are rigidly maintained (Barbour 1999, 21–22).

*Independence from a scientific standpoint*

Jay Gould, a twentieth-century Harvard University evolutionary biologist and popular writer, developed the position he called nonoverlapping magisteria (NOMA). His position is described in his book *Rock of Ages: Science and religion in the Fullness of Life* and his article "Non-overlapping magisteria" in the periodical *Natural History*, which was a response to Pope John Paul II's October 1996 address on evolution to the Pontifical Academy of Sciences. For Gould, a magisterium is a domain of teaching authority. The magisterium of science exercises its authority over the empirical realm while that of religion covers "questions of ultimate meaning and moral value." For best results, the two areas should stay within their own bounds of competence (Gould 1997; Gould 1999, 6).

Gould discusses historical instances in which religious figures overstepped their areas of competence into science, such as during the Galileo and Arkansas creation trials. He also criticizes scientists who attempt to extrapolate scientific findings into broader theological, philosophical, or ethical conclusions. For example, he criticizes Wilson's *Sociobiology* for attempting to ground moral human decision making in deterministic genes and evolution. Interestingly, Barbour points out that Gould oversteps his own bounds by making statements that appear to be supported by science but in fact should be viewed as naturalistic philosophical interpretations, involving language such as "cosmic insignificance" and "sublime indifference of nature." Barbour quotes Gould: "[Mankind is] a wildly improbable evolutionary event, and not the nub of the universal purpose . . . We are the offspring of history, and must establish our own paths in this most diverse and interesting of conceivable universes—one indifferent to our suffering, and therefore offering us maximal freedom to thrive, or to fail, in our own chosen way" (Barbour 1999, 100; Gould 1999, 206–207).

*Independence from a religious standpoint*

Since Protestant neoorthodoxy maintains that God acts in human history especially in the person of Jesus, rather than in the natural world, it easily accepts evolutionary biology. According to neoorthodoxy, the doctrine of creation is an affirmation of dependence on God and the essential goodness and orderliness of the world rather than a theory about beginnings or the natural processes that developed. Neoorthodoxy takes an informed literal approach to scripture as described in chapter 1. There are two major aspects of neoorthodoxy that allow it to maintain an independence position. One is its strong emphasis on God's transcendence and a minimization of his immanence, which leads to a gulf between God and nature. The other is that neoorthodoxy strongly dichotomizes human and nonhuman nature. Thus, neoorthodoxy leaves little wiggle room for continuing creation (Barbour 1999, 100–101).

Catholic thinking, following Thomas Aquinas, has held that God, as primary cause, works through secondary causes, which science investigates. Since these levels of cause are qualitatively different, the scientific process and the theological process may be maintained as two separate, nonrelating processes. The scientific account is complete in its own level of causation without gaps in which the primary causation would need to intervene. At the same time, theologians can hold that God sustains and makes use of this entire sequence of causation. Thus, theologians, dealing with primary causation, and scientists, with secondary causation, answer very different questions (Barbour, 101–102)

William Stoeger, a Jesuit astronomer, takes the position that God's purposes are built into nature's potentials to change and develop and that God also sustains the system from moment to moment, maintaining it in being. Without God, nature would cease to exist (primary cause). So God acts through the laws of nature (secondary cause), fulfilling his intended goals. Stoeger writes, "If we put this in an evolutionary context . . . we can conceive of God's continuing creative action as being realized through the *natural unfolding* [emphasis mine] of nature's potentialities and the continuing emergence of novelty, of self-organization, of life, of mind and spirit." In his position, Stoeger maintains that we must respect the integrity of the created order accessible to science and, thus, the integrity of science. No gaps or special divine interventions were required for the appearance of life or human consciousness—these events were results of *natural* unfoldings. He also maintains that creation of human persons with human consciousness was central to God's purposes, even though they evolved by *natural* means. However, Stoeger adds that God can use distinctive means of self-revelation to persons, such as through the person of Christ. So in Stoeger's independence position, he maintains a position on nature and evolution that relegates God's activity to remote primary causation, maintaining that God's influence ends where nature's unfolding processes begins (secondary causation; science). He leaves open the possibility of God's direct communication of significant information in people's lives (religion) and that God maintains the existence of the universe in time.

Thus, Stoeger maintains the two separate domains of science and religion, each the object of a "separate magisterium" (Barbour 1999, 102; Stoeger 1995, 249), with two caveats: (1) that God directly communicates in peoples' lives through information and (2) that God directly maintains existence in time. In separating natural processes from God's activity as carefully as he does, Stoeger's position approaches the Deist position with respect to the universe other than human spiritual life. Deism is a philosophical position that maintains the existence of a god on the basis of reason and observation of nature alone. Deists hold that God is a sort of architect of the universe. The architect doesn't alter the plan by intervening in human affairs or by suspending or interfering with the natural laws of the universe. God's activity gets it all started, after which nature takes off on its own, a bit like a windup toy.

## Dialogue

In contrast to independence positions, dialogue positions focus on similarities between scientific and religious knowledge rather than on their differences as independence positions do. Major areas of dialogue concern presuppositions in scientific and religious thinking, limit questions, similarities of methods, and analogous concepts held by scientific and religious thinkers (Barbour 1999, 23–27).

## Judeo-Christian presuppositions and the rise of science in the West

Why did modern empirical science rise in the Judeo-Christian West? It appears that the specifically Jewish idea of creation gave a religious legitimacy to taking a matter seriously without worshipping it as required for the development of empirical science. How do the Genesis creation narratives support this notion? Like the Greek philosophers, the biblical writers asserted that the world is orderly and intelligible. However, the Greeks took the position that the world's order is necessary so that one can rationally deduce its specific structure from principles. The biblical writers were unique in maintaining that God created not just matter but also matter's form, or structure. The implication is that the world doesn't have to be just as it is but that the details of its order can only become known by observation and not by reason alone. Also, since in the biblical view the world is real and good, but not divine as other ancient cultures held, it is therefore worthy of investigation. Experimentation on it is acceptable. However valuable the Jewish tradition—inherited by Christianity—may have been to the rise of science, it is also important to recognize that Arab science, especially in the Middle Ages, is an important contribution to the world's scientific heritage (Barbour 1999, 23). Eastern religious traditions tend to hold that matter is less real, even illusory, and suspect. For example, *Maya* is a Vedanta Hindu concept that maintains that the soul and the body are very different entities. The body is matter. Matter entangles and entraps the soul. In seeking to control and enjoy matter, the soul is led into illusion and away from truth and enlightenment. Maya refers to this concept of the material as illusory. *Illusion*, in this sense, refers to the effect that matter has on the soul (Dasa 2004).

## Limit questions

Limit questions are questions that can be asked by science but cannot be answered specifically within science. For example, scientific findings reveal order in the universe that is rational and contingent. In other words, the rational laws and initial conditions of the universe were not necessary (note the congruence with Jewish thought). The fact that the universe and its intelligibility are contingent inspires scientists to search for new and unexpected forms of rationally accessible order. The question, from where does this contingent but rational order come? arises. Theologians, of course, hold that God is the creative ground of this contingent and rational order of the universe. According to Thomas Torrance, a Scottish theologian, "Correlation with that rationality [of contingent order] in God . . . explains the profound sense of religious awe it calls forth from us." He refers to Albert Einstein's words that "religious awe is the mainspring of science" (Barbour 1999, 24).

Another obvious example of a limit question came to the forefront of awareness with the discovery of the big bang theory by the Belgian Jesuit priest-physicist Georges LeMaître. The theory brought into sharper awareness the temporal and spatial boundaries of the universe, as well as its contingency. The new awareness in turn brought the question, why is there a universe at all? into sharper focus (Barbour 1999, 24).

*Similarities of methods and analogous concepts*

Historians and philosophers of science—as well as, inevitably, theologians—have taken note of a sharp contrast often drawn between science as objective and religion as subjective. Barbour maintains that there are differences in emphasis along these lines between the two areas of inquiry but not in a black-and-white fashion. For example, theoretical assumptions influence the selection, reporting, and interpretation of data. Thus, data selection is not entirely theory-free. Theories don't arise directly from logical analysis of data but from activity that involves creative imagination in tandem with analytical thinking in which analogies and models play a role. For example, conceptual models (e.g., the evolutionary tree of life) help us to imagine what isn't directly observable (Barbour 1999, 25).

Of course, these same characteristics are also found in religious thinking and reasoning, only more so. The data of religion include religious experiences, rituals, scriptural texts, and historical data, which are full of conceptual interpretations (e.g., acts of God, fulfillment of prophecies). Unlike scientific hypotheses, religious belief is not accessible to empirical testing. However, beliefs can be approached with a similar spirit of inquiry to that of science (Barbour 1999, 25). For example, take the method used in chapter 1, in which the question was asked whether the Genesis creation narratives are compatible with evolutionary theory. Biblical scholars used a method of investigation and empirically based fact finding and deduced an informed literal understanding of these creation narratives. Conclusions reached were that (1) the cosmological and biological statements in these stories were based on earlier ancient Near Eastern traditions and not meant to be taken word-for-word in the absence of knowledge of those traditions

JOSEPH FORTIER

and (2) firm faith statements are made, which are original to Hebrew faith tradition, not found in earlier ancient Near Eastern traditions, and laden with value and ultimate meaning.

## Information and science-religion dialogue

Information is a fertile ground for science-religion dialogue, as it is vital to understanding complex systems accessible to scientific observation yet, in itself, is difficult to define using purely scientific concepts. Information occurs in a system in which ordered patterns may emerge. Haught defines *information* as "the overall ordering of entities—atoms, molecules, cells, genes, etc.—into intelligible forms or arrangements" (Haught 2000, 70). One might also add language to this list—the overall ordering of sounds or written symbols into intelligible forms or arrangements. A specific unit of information is one among many possible sequences or arrangements of entities in such a system. Examples of such systems are binary digits used by computers to process information, an alphabet, DNA bases, and amino acid sequences in proteins. Information is communicated when another system selectively responds to only certain patterned arrangements of components. Examples of systems that can respond to information are other computers, human readers, living cells. The meaning of the message depends on a wider context in which the information can be interpreted. Thus, information can only be viewed as an active, relational process rather than as a closed system unto the pattern itself (Barbour 1999, 105–106). With respect to physical phenomena, information may present itself as a disjunction. An example I used in a class I taught went like this. I scribbled with chalk randomly on the board, then purposefully wrote the name of a student:

I told the students that the chalk and the blackboard were two kinds of matter that could be analyzed with physics and chemistry. Furthermore, the *interaction* between the blackboard and the chalk sticking to it was similarly explainable. I watched their faces as I scribbled. Nothing. When I wrote *Erin*, suddenly they were expressive and looked at a woman in the class named Erin, who suddenly had a self-conscious, somewhat embarrassed smile.

Then we discussed the completely different kind of relationship that had emerged between the blackboard and us when I finished the *name Erin*. A unit of information composed of entities in an alphabetical system I wrote on the board (someone's name) had emerged, while nothing had changed with respect to the relationship between the chalk and the blackboard. At the onset of writing the *E* in Erin's name, a sudden disjunction with what had preceded the *E* occurred. Information started to emerge within the dynamic relational system of the class. The segment with the student's name was qualitatively different from the scribble. I conveyed information to the class that created awareness of a particular student, causing everyone (including the student, embarrassed but amused) to be aware of her. I used the chalk and blackboard as physical conduits of the information I intended to transmit. I used the alphabetical system we all understood as the informational system from which I selected a specific sequence consisting of selected informational entities (letters), out of a huge multitude of possible sequences, to symbolize the specifically intended information. Only the segment of the chalk line with the alphabetical code was decoded and reacted to by the students since only that segment transmitted meaningful and relevant information to them in their context. Yet continuity remained with the physical and chemical laws governing the behavior of chalk and blackboard—no disjunction there. The discontinuity of emergent information occurred in a quiet, unobtrusive way with respect to the physics and chemistry of chalk and blackboard.

Barbour holds that information is communicated when another system responds selectively, that is, when information is coded, transmitted, and decoded. The meaning embedded in the message depends on a wider context of interpretation. In the above case, I was the coder. I transmitted the coded information via the chalk and blackboard. The students were the "other systems" who selectively decoded and responded to the segment of the chalk line with information, interpreting it in the wider context of the relational system of the class. Thus, the informational class exercise was dynamic and relational (Barbour 1999, 106).

Similarly, in biological systems or organisms, information is carried in the base sequences of DNA as was discussed in chapter 3. Any order of the four bases, the entities A, C, G, and T, is theoretically possible in a

DNA molecule. However, the order of these bases in the DNA molecules of a particular living organism is only intelligible in the organism's context. A specific base sequence codes for a specific sequence of amino acids in a protein (chapter 3), which in turn has a specific function within the relational context of the myriad of functions that define the organism. If at any stage, the ordering of these DNA base or protein amino acid sequences is altered sufficiently, the protein will not function and the organism's life will be threatened. In fact, two types of mutations are caused by what are called "missense" and "nonsense" sequences. Life depends on relevant genetic information to make sense—be accurately coded, replicated, transmitted, and decoded—as surely as coherent human social functioning does. A review of chapter 3 will make clear that information-laden coding is present in the base sequence of DNA; the base sequence is coded information. Transmission of coding occurs during cell division (mitosis) and gamete formation (meiosis), during which DNA copies are passed on to other cells from the mother cell. Transmission of coding also occurs when the protein RNA polymerase transcribes the base sequence of a gene on the DNA molecule into a base sequence of RNA. Decoding occurs when the RNA base sequence is translated in the cell into a corresponding amino acid sequence of a protein, which can perform a specific function in the context of what the organism needs in its environment.

An analogous religious concept to information in the Christian tradition is that of divine Word (Logos). Theologians such as John Polkinghorne have proposed that God communicates information in a way that does not require violation of scientific laws. This notion of Logos can be interpreted as God's communication of rational structure and meaning when the world is viewed in a broad context. Thus, these theologians open dialogue with science by seeing science not as a source of proofs for God's existence but rather as a source of new analogies for discussing God (Barbour 1999, 61).

Haught describes a similar religious concept that comes from Taoism in the East, yet he is careful to state that he isn't implying anything mystical or supernatural. In Taoism, the ultimate reality, or Tao, is thought of as energetically passive but informationally active. The way Tao works in the world is called *wu wei*, a sort of "non-interfering effectiveness" (Haught 2000, 78–80). Haught writes that information is "quietly resident in nature" and that in spite of not having energy or mass that is accessible to empirical measurement, it is a powerful source of patterning natural elements and processes into "hierarchically distinct domains" (analogous to the sort of disjunction that I showed my students with the chalk and chalkboard) such as living biological systems, social organization in some animals, and the capacity to learn and communicate information with abstract signals in others (Haught 2000, 70–71). Haught sees information as an aspect of nature yet in itself not an easily accessible one to scientific analysis. He describes information as "insinuating itself" into the universe without violating physical or chemical laws and making use of physicochemical processes in its ordering activity while not interrupting these processes.

The topic of information opens potentially exciting and mutually informative possibilities for dialogue between scientists and theologians. It is a vital concept in both fields, is dynamic and relational while maintaining a sort of unobtrusive reserve, which lends it a tantalizing je ne sais quoi that begs exploration. It plays a large role in the process of biological evolution in emergence of disjunct hierarchies such as emergence of living organisms from nonliving matter, emergence of living forms with the capacity to develop adaptive behavior by cognitive learning from living forms without such an ability.

## Integration

The fourth category in Barbour's fourfold typology of how people relate science and religion is closely related to the dialogue category but carries it to its logical end, which is to develop an integration of the two disciplines. The value of an integrated position is that the proponents find an articulation of a mutually agreed upon, coherent vision of reality that does not compromise either discipline and affirms the value of both in describing a coherent view of reality.

Two major avenues within theology that attempt to find this integration are *natural theology* and a *theology of nature*. The two approaches differ in their starting points. While natural theology begins with the findings and observations of science as it focuses its gaze on the natural world and universe, a theology of nature begins with a concrete religious tradition based on historical experience and historical revelation (Barbour 1999, 28–31). Let us glimpse at how these two approaches play out.

### Natural Theology

Natural theology holds that we can learn about God by observing nature. In its older form, its adherents used it to attempt to prove the existence of God. Isaac Newton wrote that the eye could not have been contrived without skill in optics. William Paley, the British clergyman and naturalist, advocated what has become known as the *argument from design* for the existence of God. He said that if one were to find a watch on the heath, he would be justified in concluding that it was designed by an intelligent being. Analogously, when one considers the human eye, in its complex intricacy of form for the single purpose of vision, the only conclusion one can come to is that the eye is a product of intelligent design (Barbour 1999, 28–29). Charles Darwin wrote in *Origin of Species* that "to suppose that the eye, with all its inimitable contrivances for adjusting the focus to different distances, for admitting different amounts of light, and for the correction of spherical and chromatic aberration, could have been formed by natural selection, seems, I freely confess, absurd in the highest possible degree" (Darwin 1859, chap. 6). However, he continues by explaining exactly how the eye may have been formed by natural selection and descent with modification. Darwin's explanation of eye development by evolution has since been verified by the modern sciences of genetics and evo-devo (evolution and development) (Yoon 1994; see chapter 4, "Evo-devo and the case for unique common ancestry of all animals."

Intelligent design, an outgrowth of this older form of natural theology, attempts to pose as science and, thus, as an alternative to biological evolution in the science classroom. It was roundly rejected as science by a US federal court *(Tammy Kitzmiller et al.. v. Dover Area School District* Case No. 04cv2688). The judge ruled that intelligent design is a form of creationism, or a religious belief that life did not come about by descent with modification under natural selection but rather by instantaneous creation by God. Thus, to teach it in public school classrooms is a violation of the First Amendment to the US Constitution.

Richard Swinburne presents a modern rendition of natural theology. He argues that in science, a theory has initial plausibility, which is strengthened or weakened by various kinds of evidence. Based on the parsimony of God's existence as an explanation for causality of the world's existence, on the evidence of order to bolster the former point, and by maintaining that science alone can't account for the presence of consciousness, Swinburne argues that total evidence suggests that God's existence (theism) is more probable than not (Barbour 1999, 29).

Another modern development in natural theology is the *anthropic principle*, which notes that astrophysicists have found that the universe seems fine-tuned for the emergence of life. This fine-tuning occurred in the first second after the big bang that initiated the universe. If the rate of the universe's expansion, says the prominent astrophysicist Stephen Hawking, had been off by one part in a hundred thousand million million (a number one with seventeen zeros behind it), then the universe as we know it would not have come into being—it would have either recollapsed or expanded so that the diversity of kinds of matter that we have would not have come about (Barbour 1999, 29). Although some interpret this data as strong evidence or proof for the existence of God, others point to the possibility of multiple universes either in space or succeeding one another in time and conclude that the peculiarities of our universe that make it life friendly are a product of chance.

The appeal of natural theology is that despite cultural and religious differences, rational, informed people the world over can be expected to agree on scientific data. This scientific data often causes the awe and wonder that scientists experience in their work. (Albert Einstein: "Religious awe is the mainspring of science.") In its modern form, natural theology doesn't pretend to prove the existence of God, but rather, it proves that it is more plausible than other alternatives (Barbour 1999, 30), which also appeal to reason.

*Theology of nature*

With respect to biological evolution, a theology of nature begins with the religious experience and life of a given religious community and holds open the door to the community's ongoing dynamic, relational spiritual growth precisely by exposing it to the secular sciences, including updated biological research findings. The door held open is an attitude of confidence that the process of absorption of new information and reflection leads to the subsequent growth of the community's own understanding of God's relationship with the world. Process thinking, as we shall see in following chapters, makes a major contribution to this attitude of openness to the process of scientific and theological discovery and integration of those discoveries taken by this perspective. This spirit of openness in theologies of nature is expressed in passages of the Catholic document *Gaudium et Spes* (*Pastoral Constitution on the Church and the Modern World*) which itself was a product of the Second Vatican Council of the Catholic Church, between 1963 and 1965—in which the leadership of the Catholic Church expressed the Church's need for information found by the modern sciences in order to more effectively understand and teach Christian faith. A passage from article 62 of *Gaudium et Spes* is especially clear in expressing this openness:

JOSEPH FORTIER

Experience shows that, for circumstantial reasons, it is sometimes difficult to harmonize culture with Christian teaching. These difficulties do not necessarily harm the life of faith, rather they can stimulate the mind to a deeper and more accurate understanding of the faith. The recent studies and findings of science, history and philosophy raise new questions which effect life and which demand new theological investigations. Furthermore, theologians, within the requirements and methods proper to theology, are invited to seek continually for more suitable ways of communicating doctrine to the men of their times; for the deposit of Faith or the truths are one thing and the manner in which they are enunciated, in the same meaning and understanding, is another. In pastoral care, sufficient use must be made not only of theological principles, but also of the findings of the secular sciences, . . . so that the faithful may be brought to live the life of faith in a more thorough and mature way.

May the faithful, therefore, live in very close union with the other men of their time and may they strive to understand perfectly their way of thinking and judging, as expressed in their culture. Let them *blend new sciences and theories and the understanding of the most recent discoveries with Christian morality and the teaching of Christian doctrine* [emphasis mine], so that their religious culture and morality may keep pace with scientific knowledge and with the constantly progressing technology. Thus they will be able to interpret and evaluate all things in a truly Christian spirit. (Abbott 1966, 268-269)

One is reminded of Anselm of Canterbury's motto: "Faith seeking understanding" (Williams 2007).

The thrust of the above passage calls upon scholars of Christian doctrine to mix with those of secular disciplines such as science to learn from them and to "blend new sciences and theories and the understanding of the most recent discoveries with . . . teaching of Christian doctrine." In short, article 62 of *Gaudium et Spes* offers a strategy for developing a practical theology of nature. Start with a living religious tradition, absorb new scientific information, reflect, and integrate it into an updated interpretation of one's faith tradition in a way that is true to the tradition.

In 1996, Roman Catholic pope John Paul II, not universally considered a church progressive, wrote an address to the Pontifical Academy of Sciences in Rome, which was a major step forward in the science-religion dialogue toward developing an integrated theology of nature. Pope Pius XI had founded the Pontifical Academy of Sciences in 1936. Its goal is to promote progress in the mathematical and natural sciences. The academy boasts an international membership of highly respected scientists, including several Nobel laureates. The title of the document, "Truth Cannot Contradict Truth," paraphrasing and honoring Galileo from an earlier time, states the reason the pope felt that the time had come for Christianity to welcome biological evolution and adjust its teaching accordingly. He wrote, "New knowledge has led to the recognition of the theory of evolution as more than a hypothesis. It is indeed remarkable that this theory has been progressively accepted by researchers, following a series of discoveries in various fields of knowledge. The convergence, neither sought nor fabricated, of the results of work that was conducted independently is in itself a significant argument in favor of this theory." The pope then reverts to an independence position, accenting the distinction between undeniable scientific finding and the teaching authority of the church and the issue of soul. Nonetheless, it is an intellectually honest independence position and invites dialogue toward integration.

In the following chapters, we'll explore the possibility of a theology of nature that would be true to biblical Christianity and would integrate modern scientific discovery and thought with Christian philosophy and theology to produce an intellectually responsible way of faith for the present scientifically informed age, specifically with respect to biological evolution. We'll start this exploration in chapter 6 with the thinking of Stuart Kauffman, whose materialist conflict position was summarized above. Although a scientific materialist philosophically, Kauffman's work with the science of complexity is immensely helpful to developing an integrationist position between Christian faith and biological evolution.

# Summary

☐ According to Ian Barbour, the positions that people take with respect to evolution and religion fall into four major categories: conflict, independence, dialogue, and integration. The conflict category is composed for those theists, atheists, and agnostics who agree that evolution and religious faith are irreconcilably opposed to each other. Of course, the theists who hold this position hold that religion is true and evolution is false, while the atheists and agnostics hold the reverse.

☐ A concern that many theists who take a conflict position hold is that if we were to believe that we are just other animals who've evolved from animal ancestors, then there is no ground for purpose, human dignity, and morality. The Muslim scholar Seyyed Hossein was concerned that loss of a sense of the sacred origins of nature would result in loss of meaning and moral sensibility, and wholesale destruction of the earth systems would result because of careless human activity.

☐ Some scientific atheists and agnostics who take a conflict position hold that religion is explainable by biological evolution and that it is a sort of vestigial trait no longer of use. These people accept the practical working philosophy of the scientific endeavor, empiricism (the only way of knowing anything is through the senses), and extrapolate this philosophy to all of life. Until Stuart Kauffman, it was generally accepted among many scientific atheists and agnostics that all explanation for living, thinking, and emotional phenomena is reducible to physical and chemical laws and principles—a position called reductionism. Kauffman, however, rejects this argument for complex-networked systems such as living systems, in which he claims that robust "order for free" emerges and condenses spontaneously when such systems are poised at the edge of chaos. The rules that govern such emergent order cannot be predicted a priori, although the molecular components of such systems remain explainable by physical and chemical laws.

☐ Adherents of independence positions seek to avoid conflict. Both religious believers and scientific agnostics and atheists hold some version of Gould's nonoverlapping magisteria, in which religionists have their legitimate claims to authority, as do scientists. For best results, these areas of authority should not overlap but, rather, should stay within their own bounds of competence. Religionists who take an independence position tend to play up the transcendence of God and to minimize the importance of science. There is a tendency to strongly dichotomize human versus nonhuman nature. Independence-position holders tend to focus on differences between scientific and religious knowledge.

☐ The dialogue positions accent the similarities between scientific and religious ways of getting knowledge and focus on ways that science and religion complement one another. For example, science finds that the universe has rational laws, has order, and is contingent. Religion asks, "From where does this order come?" Both science and religion are interested in information. It is vital to understanding complex systems such as those Kauffman investigates, but is difficult to define using scientific concepts. Information is carried in the genetic material of living organisms. Religious ideas such as the Christian divine Logos and the Taoist *wu wei* are similar ideas to information.

☐ The fourth category in Barbour's fourfold typology of how people take positions on science and religion is integration. Integrationists seek to develop positions that integrate the two disciplines. Two major approaches to integration are natural theology and theology of nature. Natural theology begins with scientific observations and attempts to find evidence for God from that starting point. In contrast, theology of nature assumes the truth underlying its own religious tradition and approaches science in order to inform itself and better interpret and articulate its tradition in light of scientific findings. Intelligent design is a development of an older, early nineteenth-century form of natural theology, which attempts to hold that complex organs and organisms are "irreducibly complex" and could only have come about by direct divine intentionality. It attempts to pose as science, but in a 2005 court case in Pennsylvania, USA, it was spectacularly discredited as such. A modern development in natural theology is the anthropic principle, which notes that astrophysicists find that the universe, from the first microseconds after the big bang singularity that initiated its development, seems fine-tuned for the development of life and human intelligence. Kauffman's position that basic laws in the universe exist and predispose it to producing intelligent life reinforce this idea.

☐ Process thought has been very helpful in developing theologies of nature, such as those developed by various Christian thinkers. Documents in the Catholic Second Vatican Council, which occurred in the early 1960s, contain calls of that church to itself to look to science and the humanities to more effectively understand its underlying faith traditions.

# References

Abbott W. M. 1966. *The Documents of Vatican II*. New York: Guild Press.

Augustine of Hippo. 1982. *The Literal Meaning of Genesis*. New York: Newman Press.

Barbour I. 1999. *When Science Meets Religion*. San Francisco, CA: HarperSanFrancisco.

British Broadcasting Service. 2008. "Pope Praises Galileo's Astronomy." Last modified 2011. http://news.bbc.co.uk/2/hi/europe/7794668.stm

Dasa R. 2004. Maya: "Illusion." In *The Heart of Hinduism*. Hertfordshire, England: ISKCON Communications. Last modified 2004. http://hinduism.iskcon.com/concepts/105.htm

Founders Ministries 2011. "The Abstract of Principles." Last modified 2011. http://www.founders.org/abstract.html

Futuyma D. J. 1995. *Science on Trial; the Case for Evolution*. Sunderland, Massachusetts, USA: Sinauer Associates, Inc.

Gould S. J. 1997. "Nonoverlapping Magisteria." *Natural History* 106: 16-22

Gould S. J. 1999. *Rock of Ages: Science and Religion in the Fullness of Life*. New York: Ballentine Books.

Haught J. 2000. *God After Darwin*. Boulder, Colorado: Westview Press.

Pope John Paul II 1996. Message to Pontifical Academy of Sciences October 22, 1996. Updated December 3, 2010. http://www.cin.org/jp2evolu.html

Kauffman S. 2008. *Reinventing the Sacred: A New View of Science, Reason, and Religion*. Philadelphia, Pennsylvania USA: Basic Books.

NewScientist. 1992. "Vatican admits Galileo was right." http://www.newscientist.com/article/mg13618460.600-vatican-admits-galileo-was-right-.html

Noll MA. 1992. *A History of Christianity in the United States and Canada*. Grand Rapids, Michigan: W. B. Eerdmans.

Owen R. and S. Delaney. 2008. "Vatican recants with a statue of Galileo." London Times March 4. Last modified 2010. http://www.timesonline.co.uk/tol/comment/faith/article3478943.ece

Pope Pius XII 1950. "Humani Generis." Last updated July 28, 2011. http://www.papalencyclicals.net/Pius12/P12HUMAN.HTM

Sandeen E. R. 1967. "Towards a Historical Interpretation of the Origins of Fundamentalism." *Church History* 36: 66-83

JOSEPH FORTIER

The Emmanuel Bible Church. 2009. Doctrinal Articles of Faith. http://www.ebcsmo.org/statementoffaith.php

Stoeger W. ed. 1995. *Describing God's Action in the World in Light of Scientific Knowledge*. Rome, Italy and Berkeley, CA: Vatican Observatory and Center for Theology and the Natural Sciences.

Unger W. 1981. *"Earnestly Contending for the Faith": The role of the Niagara Bible conference in the Emergence of American Fundamentalism, 1875-1900*. Simon Fraser University, Burnaby, British Columbia, Canada.

Williams T. 2000. "Saint Anselm." Latest modification 2007. http://plato.stanford.edu/entries/anselm/

Wilson EO. 1975. *Sociobiology: the New Synthesis*. Cambridge, Massachusetts: Harvard University Press.

Wilson EO. 1978. *On Human Nature*. Cambridge, Massachussets: Harvard University Press.

Wilson EO. 1995. *Naturalist*. New York: Grand Central Publishing.

Yoon C. K. 1994. "The Wizard of Eyes: Evolution Creates Novelty by Varying the Same Old Tricks." New York Times November 1.

# CHAPTER 6

## Stuart Kauffman and the Science of Complexity

### After the Enlightenment and
### Darwin's bombshell, a need to reinvent the sacred

IN THE OPENING pages of his book *At Home in the Universe: the Search for the Laws of Self-Organization and Complexity*, Stuart Kauffman writes that due to the discoveries of science in the past five hundred years, we have lost our sense of our place in the universe. Kauffman, a member of the Santa Fe Institute (www.santafe.edu) during the 1990s, presently on the faculty of the University of Calgary, and MacArthur Fellowship (Genius Award) recipient, is considered to be the leading thinker on self-organization and the science of complexity as applied to biology. Before the Renaissance and Age of Enlightenment, Kauffman writes, the Western world felt secure in its place as chosen of God, at the center of the universe, created because of his love for us. The first major blow from science to this view came from Copernicus, whose astronomical calculations informed him that the earth goes around the sun and not vice versa. Later, Galileo and Keppler corroborated these findings. Why such a blow to the religious view? Kauffman writes that the geocentric (earth-centered) view of the universe wasn't just a matter of science. It was powerful evidence supporting a view of the universe that revolved around our species. After all, our species was believed to be the reason for the existence of all the rest. Church leaders feared that a compromise of this view would corrode the unity of a tradition of human rights and duties and moral fabric built over a thousand years (Kauffman, 1995, 5–6). Still, if the heavenly bodies still obeyed eternal laws, no matter to which far-flung edge of the universe Earth was eventually flung, and if the living things including man had inhabited the earth since the creation of the universe just as we see them today, then we still lived in a universe that was not completely out of touch with the medieval Christian worldview; so the moral and human rights fabric of our society was still basically intact.

Then along came Charles Darwin. The former innocently anthropocentric view of nature would never be the same again, except perhaps in obscure corners of human awareness. Many present-day neo-Darwinists see the origins of humanity as the result of a chain of random accidents, including chance mutations, each contingent on the former, sifted by the law of survival of the fittest, in a living world governed more by a meaningless sequence of environmental change and its effects on genes than by any purpose. Others are not quite so sanguine about this scenario. The science in Christian "creation science" is not science at all, Kauffman notes, but he respects the moral anguish and sense of loss of the sacred that underlies creation science (Kauffman 1995, 6). As we saw in chapter 5, Seyyed Hossein Nasr, the distinguished Muslim scientist and religious thinker, eloquently expressed these sentiments.

Kauffman asks whether we need to replace our traditional creation myths with a creation myth that includes a nonreductionist, scientific approach to the emergence of life (Kauffman 1995, 54). I submit that nothing needs to be replaced or added to the Genesis creation myth; it stands by itself in its religious assertion that a moral, relational god is ultimately the creator and ground of existence, that the material universe is both real and good, and that the human person is dignified in the image of God. In a similar way, the American Indian worldview, which reports hierophanies or manifestations of the sacred (powers) in various natural creatures, needs also to be taken more seriously than simply relegated to the catacombs of superstition. The native people of the New World had a unique, intimate relationship with the natural world. Their experiences informed them that animals and other natural entities mediated sacred presence with an interest in care, value, and dignity that bespoke important similarities with other great religions. They experienced certain natural phenomena not as deities but as servants or intermediaries between humans and, in various cultures, the Creator, the Powers, the Great Mystery. They, as the Hebrews, experienced the natural world as real and were confident in its goodness. They also experienced the Powers, Creator, or Great Mystery as benevolent and sought aid from that source. We simply need to understand our religious narratives in their cultural, historical, and spiritual context and also come to understand them in the context of the discoveries of modern science after the methods of a theology of nature. With respect to the Genesis creation narratives, what is needed is a true understanding of how they represent the depths of Judeo-Christian spiritual heritage regarding God's relationship with the universe and with humanity and an accurate and sufficiently well-informed grasp of the findings of the science of evolution. The science of complexity and the evidence in support of it described by

Kauffman and others go a long way in making this integration possible. As we shall see, the science of complexity offers itself as a meeting place at which evolutionary biology and religious faith in the transcendent god can meet and find the possibility of integration. In chapter 7, this possibility presented by complexity will be explored further.

## Introduction to the science of complexity

The theory of natural selection presumes contingent random events determining other contingent random events as *the only source* of the tremendous order and pattern we observe in life and living systems. Selective forces in the environment act on random genetic variation in a given population of organisms. And just as those forces of natural selection in the environment occur randomly within the environment over time, those genetic variations in the population of organisms were *themselves* caused by accidental genetic events. Kauffman takes issue with these natural mechanisms as the *sole cause* of descent with modification and the rise of adaptive order (wings from forelegs from fins; a brain capable of reason arising from a simple nerve ganglion). For Kauffman, the emergent sciences of complexity suggest that order doesn't emerge from the random, contingent factors upon which natural selection depends. Rather, those selective factors work on order that arises from a wholly different source, as well as *along with* this more primal source of order, as new order emerges. If there are basic physical laws in the universe that underlie the emergence of order upon which natural selection works, then complex living organisms (i.e., *all* organisms) are not merely tinkered-together products of accident and chance but, rather, are to be expected to arise within our universe (Kauffman 1995, 8), although precise details of final outcomes may never be predicted a priori (Kauffman 1995, 17–18).

Kauffman doesn't just agree with evolutionary biologists Stephen Jay Gould and Niles Eldredge, the authors of the theory of punctuated equilibrium. He explains what may well be the mechanism behind what is observed in the fossil record. According to the theory of punctuated equilibrium (punk eek), contrary to Darwin's notion that the evolutionary process of biology is gradual and constant, change instead occurs in jumps that follow long periods of stasis (punctuated equilibrium). Kauffman explains that this is what one would expect from the behavior of matter and energy from the viewpoint of the science of complexity.

JOSEPH FORTIER

## The Cambrian explosion

For example, the fossil record informs us that life began gradually. The first evidence of bacterial cells, from fossils, has been found in 3.45-billion-year-old rock. Then, for almost 3 billion years, there was little change. Nothing but single-celled life. Not until about 635 million years ago does the first evidence of multicellular life show up in the fossil record—scanty at that. Suddenly about 542 million years ago, a veritable explosion of diversity and complexity of various animal forms emerged: over about 12 million years. For us, 12 million years is not sudden. In the vastness of geological time, it is. During the Cambrian period (see chapter 4, geological timescale), every known animal phylum, or major group with a distinctive body plan, emerged—plus some that seem very strange by modern standards, which soon became extinct. Over the next 100 million years, the fossil record shows that the original ancestral populations of each phylum radiated into many new species. Within clusters of similar species, related by common ancestry, new buds would branch; and subclusters of similar species, each sharing unique common ancestry not shared with descendents of other buds, would radiate (Kauffman 1995, 10–13). Thus, from ancestral wormlike creatures whose fossils are known as *Pikaia*, with notochords resembling larval sea squirts (National Museum of Natural History 2009), a diversity of jawless filter-feeding fishes arose. From within a cluster of these jawless fishes, a bud population of fish with jaws first arose, which in turn gave rise to the ancestral populations of all modern fish as well as land animals (see chapter 4, "Is there evidence of descent with modification in the fossil record?").

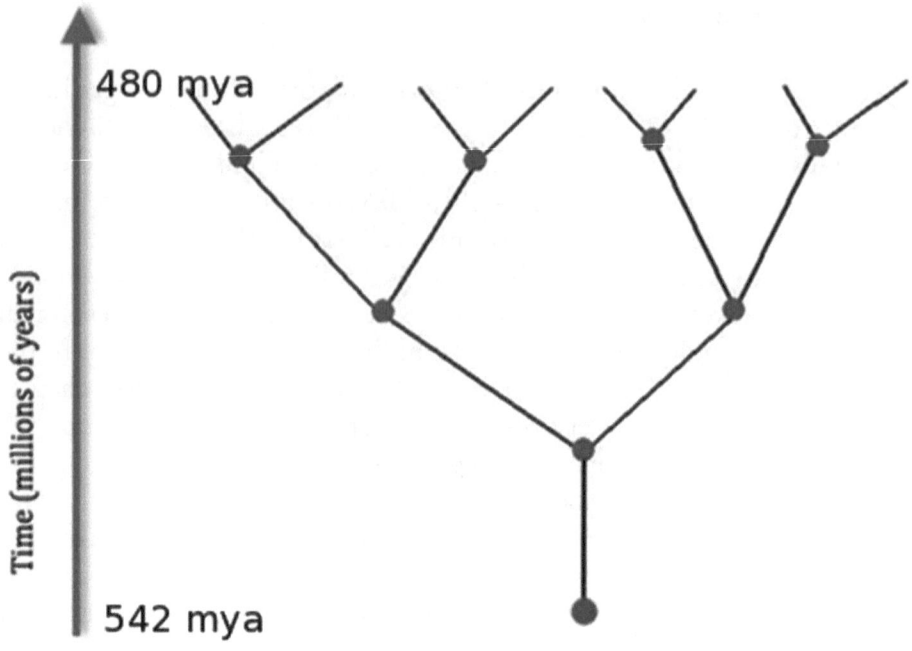

**Figure 6.1. Gradual evolution. Assuming Darwin's idea that evolution always happens at the same gradual rate, we would expect a graph of the rate of emergence of new major groups (phyla) of organisms to always look like the above. Balls represent evolutionary nodes with ancestral populations (phyla) of organisms below them and descendent populations (phyla) of organisms above them. Each line represents the time period that a phylum persists before branching into new kinds of organisms—new descendent phyla. Notice that the distance (time) between phyla is always about the same.**

So what? "So what" is that this discovery of the Cambrian explosion turned the traditional gradualist approach of classical Darwinism on its head. If life arose at a gradual, steady rate throughout its history, then the earliest multicellular animal populations should have budded and differentiated gradually from one another, as in figure 1.

But the fossil evidence during the time between 600 million years ago and 480 million years ago is more like this:

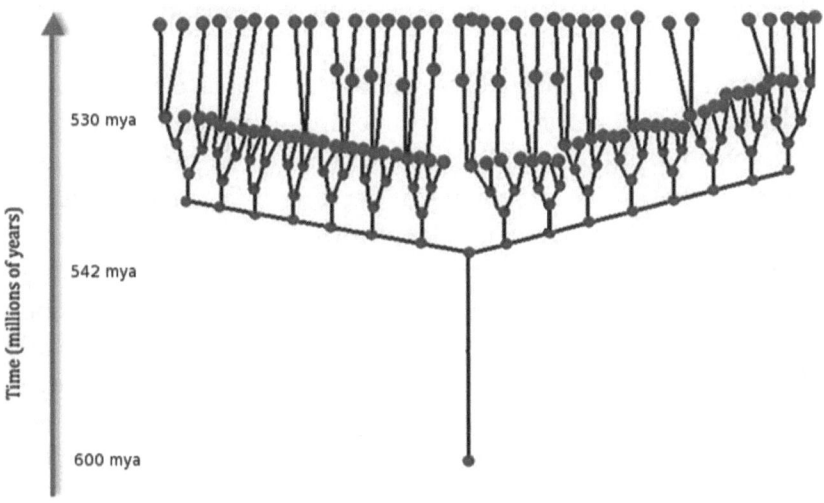

**Figure 6.2. Punctuated equilibrium and the Cambrian explosion: what actually happened according to fossil evidence.** Instead of gradual radiation of major groups (phyla) of multicellular animals, there was a sudden explosive radiation of widely various body plans, each representing a different phylum, within some 12 million years rather than hundreds of millions of years. Some were quite bizarre. The balls along the base of the radiation represent all major animal phyla, each of which possessed a radically different body plan from any other such population. Successive levels of balls at ends of branches above these ancestral populations represent radiations that inherited these respective body plans from the ancestral population, and thus are members of each given phylum represented by the lower (phylum) ball from which they descended. Some of these ancient phyla have gone extinct. All of today's 36 existing animal phyla can be traced to the Cambrian explosion—the lowest layer of balls in this diagram.

The chart of life's diversity could have been filled in from the bottom up, as in figure 1. In this scenario, multicellular life would have started with the emergence of a very simple animal phylum. The ancestral population would have diversified into various species, and one or two of those species eventually would itself have given rise to the ancestral populations of one or two other species that over generations had accumulated the genetic seeds of radically new body designs. The descendants of these species would give rise to a small number of new phyla. This scenario would have repeated this theme of gradual emergence of new phyla emerging from ancestral phyla over vast amounts of time from ancestral species all the way up to the present.

Although the above scenario is perfectly consonant with Darwinian gradualism—the notion that evolution occurs at a slow, steady rate—it is not what happened. Instead, as we saw, the fossil evidence all shows that relative gradualism of the first three billion years of life's history on earth produced a sudden, explosive emergence of all major animal phyla that we still have today, plus many more that have gone extinct (figure 6.2). The explosion resulted in a top-down, chart-filling scenario. The animal populations that suddenly emerged during the Cambrian period each represented a *distinct phylum*; a *unique* body plan. *Suddenly* (about 12 million years). This is the difference. During the next hundreds of millions of years, the chart of life's diversity was filled in from the top down: the ancestral population of each phylum diversified into species that retained the specific body plan of the ancestor right down to the present. Thus, the broadest, most general, top categories of animal life, the phyla (such as arthropods) emerged suddenly; and lower, more specific categories (such as crabs, lobsters, spiders, insects—all animal groups within the arthropod phylum) emerged later from that first arthropod common ancestral population, which gave rise to still lower, more specific categories (e.g., butterflies, dragonflies, wasps, and cockroaches are all animal groups within the Insecta).

During the subsequent course of the natural history of life on earth, there have been similar episodes of sudden bursts of emergence of new forms as well as spasms of mass extinction. We know from fossil evidence of five major mass-extinction spasms, including that of the dinosaurs about 65 million years ago, which opened ecological opportunities or niches for the rapid expansion and diversity of mammals (animals with a backbone, hair, and in which females have mammary glands and suckle their young) (Futuyma 2005, 98, 103, 104, 144, 148–149). We also know of many smaller extinction spasms interspersed among these very large ones. Extinction spasms have gone hand in hand with bursts of evolutionary creativity. We've glimpsed the creative burst that occurred during the Cambrian and the mass extinction that took out the dinosaurs (except birds, direct descendants of other dinosaurs). In fact, one of the five major extinction spasms occurred at the end of the Cambrian. During yet another major episode, dinosaur diversity emerged and blossomed after perhaps the largest extinction spasm in earth's history: at the end of the Permian period, about 290 million years ago (Futuyma 2005, 103–109).

Is there a pattern behind these bursts of creativity and bursts of extinction that the fossil record reveals? Kauffman says yes. There is evidence for patterns of emergent "order for free" from computerized modeling of complex, dynamic, adaptive systems such as living systems, ecosystems, and economic systems. This emergent order suggests deeper physical laws at work in the universe that govern complex systems. These deeper laws may not affect the particular, random, contingent accidents that cause the products of evolution to be precisely as they turn out to be but, rather, affect the overall emergent patterns of radically different novelty. For example, the particular branching of particular life-forms that emerged during the Cambrian period might well differ should the Cambrian period be rerun. But the *patterns* of branching, prolific at first, then dwindling to tweaking details later, is likely to be lawful. Thus, biological evolution, as well as the deeply historical process, is likely to be lawlike, following the twisting road of chance, contingency, and accidental events (natural selection) toward the emergence of specific forms that Darwin describes (Kauffman 1995, 14). The science of complexity explains from whence comes the order that natural selection tweaks toward the final details.

*Self-organized criticality and the edge of chaos*

Danish physicist Per Bak and others have described self-organized criticality, a close conceptual cousin of the science of complexity. Bak uses the analogy of a sandpile to describe self-organized criticality. Sand is added at a constant slow rate to the sandpile over time. Eventually, avalanches begin. What Bak has found is that avalanches follow a pattern similar to that revealed by fossil evidence of mass extinctions and periods of rapid emergence of new species. There are a lot of small avalanches and a few large ones. The addition of another grain of sand, no matter how small, can be the factor that unleashes a major avalanche or a minor one. As with mass extinctions and periods of rapid emergence of new species, we can make general statements such as that there will be more small avalanches than large ones. But in all these cases, there is no way to predict whether a particular sandpile event will be a small or a major avalanche (Kauffman 1995, 28–29) or which particular grain of sand will be the trigger.

Poised at the edge of chaos. This is how Kauffman describes complex interrelating adaptive systems, whether they be microbes, elephants, ecosystems, or lineages of organisms in their processes of evolutionary dynamism. There is an endogenous self-organization to these systems, expressed in the bubbling over into condensed emergent order, which suggests some deep, hidden lawfulness—spontaneous emergence of novel order, irreducible to what preceded the emergence of this order, the novel order with its own set of rules. There is strong suggestion that all complex interrelating adaptive systems evolve to a natural state at the edge of order and chaos, a place where structure exists but is not crystallized, a place where novelty and surprise can emerge spontaneously. At this state, poised at the edge of chaos, small and large avalanches of coevolutionary change, in which organisms and entities affect one another's evolutionary trajectory, reverberate through systems as the actors make small, best choices for themselves, interrelating to survive (Kauffman 1995, 15).

A whirlpool in a river is a good way to begin visualizing how a complex open system behaves. The whirlpool is an open system, distinguishable from the larger, more closed river system by various local factors that cause the water to swirl at that place. A system is open when it is partially bounded from its sources of matter and energy. Closed systems, in contrast, contain all the matter and energy that is available to them. The water in the river at large is a chaotic system, or nearly chaotic. The currents constantly change with respect to each other from moment to moment. The water, a substance composed of matter (water molecules), is accompanied by moving energy, which exerts force. As the chaotic water moves into the zone of the whirlpool, its motion is suddenly organized into a vortex. Although water and energy constantly move through this vortex, they are constantly replaced by more water and energy, so the vortex remains. The vortex is a pattern that maintains a robust order in which matter and energy constantly pass through. So it is with complex, dynamic, adaptive open systems like living organisms. In addition, these complex living systems are nonequilibrium systems. Varying amounts of matter and energy are constantly flowing through them since they relate with other similar open nonequilibrium systems. Thus they lack the stability of a simpler open system, such as a whirlpool.

*Computers and nonequilibrium systems*

Kauffman uses the example of a computer algorithm to explain the behavior of these complex nonequilibrium systems. An algorithm is a set of rules for solving a problem in a finite number of steps. An example of an algorithm might be the set of rules for finding the least common denominator for a set of fractions. For most deep mathematical theorems accessible to computation, in by far the majority of cases, there are no shortcuts to predicting what an algorithm will do. One must simply

JOSEPH FORTIER

execute the algorithm and observe the succession of actions and states as they unfold. The algorithm itself is its own shortest description. Thus, a computer working through a complex algorithm, plugged into an electric socket, manipulating electronic bits in silicon chips in various patterns, qualifies as an open nonequilibrium system. The theory of computation informs us that this device may be operating in such a way that it is its own shortest description. So the shortest way to predict what this physical complex system will do is to just watch it. It is possible that cells, microorganisms, wolves, ecosystems, and the earth system are also their own shortest descriptions (Kauffman 1995, 21–23).

## Reductionism called into question

The evidence suggests that complex, open nonequilibrium systems, such as living systems, are their own shortest descriptions. Thus, the assumption that it is possible to reduce all behaviors of biological entities or any other complex nonequilibrium system to physical and chemical explanations is called into question. Three other difficulties bring this assumption into focus. First, the quantum theory precludes detailed prediction of molecular behavior since much of that behavior is random. Second, chaos theory, used to study weather patterns and similar ephemeral patterns that show no stability over time, shows that minute changes in initial conditions can lead to snowballing effects of behavior in these chaotic systems. Thus, a butterfly beating its wings in Japan can be the primary cause of a tornado in Kansas as the wing-beat effect on air pressure reverberates and is magnified through the atmospheric system. And now, thirdly, computation theory seems to imply that living systems and other open nonequilibrium systems such as economic systems are similar to computers in that for by far the majority of computer algorithms, no compact, lawlike description of their behavior can be found. A reason to hope that the emergence of life and its process of evolution in which "endless forms most beautiful" constantly emerge may be governed by deep, beautiful laws governing its unpredictable, incompressible flow is because many features of organisms and of their evolutionary history are profoundly robust and insensitive to details, details such as the sorts of contingent accidents that evolutionary biologists have up to now nearly exclusively focused on (Kauffman 1995, 22–23).

Nobel laureate physicist Robert Laughlin has argued that reductionism alone is insufficient not only for complex nonequilibrium systems but also for simple equilibrium systems, such as gas particles. Think of a room filled with air. The air in the room is composed of billions of tiny invisible molecules, all shooting around and bouncing off each other at incredible speeds—as are all gases at temperatures to which we are accustomed. It is known that temperature corresponds to the average motion energy of these tiny particles. So the faster the particles are moving, on average, the higher will be the temperature. Point one, says Laughlin, is that speaking of the temperature of *one gas particle* makes no sense. Only an aggregate of gas particles displays the property of temperature. In addition, the larger the number of gas particles in equilibrium in a system (such as in a room), the more precise is any measurement of temperature. Thus, Laughlin holds that temperature emerges as a collective emergent property of the *entire* gas system and that this property is not present in any given constituent particle but *only in the whole*. Laughlin provides a number of other examples of collective emergent properties using various substances (Kauffman 2008, 24–25). In his book *Reinventing the Sacred*, Kauffman refers to these properties of systems that only emerge with the wholeness of the system as examples of "an organizational law of the Laughlin type" that depend on mathematics rather than physics (Kauffman 2008, 62).

*The emergence of life and chemical autocatalytic networks*

There is a growing body of evidence that the first single-celled living creatures emerged whole, possibly quite suddenly, in Cambrian explosion fashion, from complex aggregations of molecules containing carbon sometime about or before 3.5 billion years ago (Futuyma 2005, 92–94; Kauffman 1995, 31–69; Kauffman 2008, 44–71). Among the tiniest and genetically simplest presently living creatures are the *Mycoplasma* bacteria, many of which are human-disease organisms, such as *M. pneumoniae*, the pneumonia pathogen. *Mycoplasma genitalium*, which infects the human genital and urinary tracts, has only 485 genes. The range of gene number among the bacteria, the genetically simplest organisms as a group, is from 485 to about 5,000 (Kimball 2009; Medscape 2004; Microbewiki 2011). Why this observed minimal complexity in a living thing? No deep answer has been provided by science as yet. Kauffman suggests an application of an organizational law of the Laughlin type: matter must reach a certain level of complexity in order for life to emerge from it. The threshold is no accident of natural selection working on random variation. Rather, it is inherent in a law of nature (Kauffman 1995, 42–43).

Kauffman hypothesizes that life emerged when a chemical autocatalytic network, perhaps along the shallows of an intertidal zone thick with various simple carbon-based molecules, built to such a critical level of complexity (Kauffman 1995, 47–69; Kauffman 2008, 44–71). The probability of a large diversity and abundance of such prebiotic, carbon-based molecules was far greater before about 2 billion years ago because there was little or no oxygen in the water or atmosphere at that time (Futuyma 2005, 92–94). Oxygen tends to inhibit chemical interactions. What is a chemical autocatalytic network? A network, of course, is a system that is an interconnected set of elements. A set of rules governs what is communicated between and among the elements and what effects various kinds of communications have on any given element receiving a given kind of communication. Since the network in question is chemical, the elements in question are molecules, the small building blocks of matter, each composed of one or more atoms. The rules are derived from chemical behavior. A catalyst is a substance that promotes a specific chemical reaction—either the formation of a molecule or its breaking apart. Finally, *autocatalytic* refers to the presence of some molecules in the network that are capable of promoting their own synthesis. The simplest autocatalytic set might look something like figure 6.3. In figure 6.3, two simple molecules, A and B can join together to form either complex molecule AB or complex molecule BA. Either complex molecule acts as the catalyst for the formation of the other complex molecule and, thus, as a catalyst for *its own* formation. For example, molecule BA catalyzes its own formation by catalyzing the formation of molecule AB since AB is the catalyst for forming BA and vice versa. This is an open system that needs a supply of molecules A and B from outside the system in order to maintain the production of molecules AB and BA. If that supply fluctuates over time from moment to moment, then the system is also a nonequilibrium system (Kauffman 1995, 48–54).

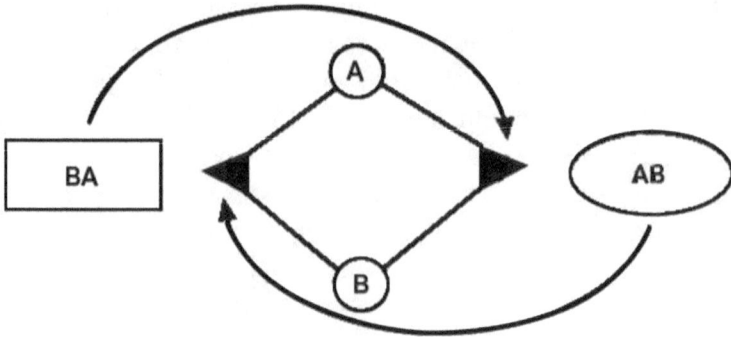

**Figure 6.3. A simple autocatalytic set. Both molecules AB and BA are formed when molecules A and B join together. Molecule BA acts as the catalyst promoting synthesis of molecule AB, which in turn is the catalyst for BA synthesis (Adapted from Kauffman 1995, 49)**

A chemical autocatalytic network is a system composed of a large diversity of both simple and complex molecules, many of which catalyze the formation of one or more others. The autocatalytic sets (such as in figure 6.3) in a chemical autocatalytic network are enmeshed with one another into an integrated nonequilibrium open network that depends on a constant supply (food source) of the various chemical ingredients in the system to remain dynamically operative. The particular patterns within the network and what regular, stable ordered patterns of chemical reaction sequences emerge over time give the system its unique properties based on these emergent patterns of chemical reactions. Emergent patterns of chemical reactions (autocatalytic networks) that compose our bodies are the approximately 256 different cell types, from liver cells to bone cells to epithelial cells that produce hair. Each cell of any given cell type is itself composed of a dynamically operative, integrated autocatalytic network of breathtaking complexity (Kauffman 1995, 48–69)

How might the emergence of the first, perhaps *Mycoplasma*-like, bacterial cell have happened? It may be that in most cases, shallow-water, anoxic (low oxygen) conditions permitted the emergence of complex, integrated autocatalytic systems up to various points of complexity. Oxygen, a powerful oxidizing agent that inhibits many chemical reactions, was not present or present in very low concentrations in the air and waters of early Earth. Then, under conditions of the "perfect storm," the process—unhampered by events that terminated the complexification of other building chemical

JOSEPH FORTIER

networks—passed a critical threshold of complexity, and the degree of patterned order necessary for sustainable life emerged. A robust network with enough vortices of self-sustaining, robust, emergent chemical order appeared, stabilized, and became integrated so that the critical amount of such order necessary to sustain a living, reproducing system emerged and persisted. This would have had to happen in a sufficiently stable environment that could provide the first living cell with a steady supply of the molecules and energy it needed in order to maintain itself, with a minimum chance of events that would disrupt the process before a robust network emerged. Life emerged from a nonliving, self-ordered, highly complex, dynamically operative integrated chemical network once a critical threshold of diversity of chemical components and network of relationships among those components was crossed.

a. *The Miller-Urey experiment*

What sorts of evidence suggest this scenario? In 1952, Harold Urey and his graduate student Stanley Miller conducted an experiment to test whether the conditions of early prebiotic Earth would produce the molecular ingredients necessary for life. They filled a glass flask with the gases methane, carbon dioxide, and ammonia—the gases that evidence suggests composed the atmosphere of early earth. Next, they sent a flurry of electrical sparks into the beaker, mimicking lightning. A few days later, brown goo had accumulated on the sides of the beaker. Analysis showed that the goo consisted of organic molecules including an impressive variety of amino acids, the building blocks of proteins. The experiment made front-page news because it demonstrated that organic molecules could be produced outside of a living organism. Subsequently, other researchers repeated the Miller-Urey experiment, varying conditions for the atmospheric composition to meet various alternate hypotheses concerning conditions of the earth. These laboratory models of early Earth have together yielded all twenty amino acids, sugars, nucleotides (building blocks of DNA and RNA), lipids, and ATP, the biological powerhouse molecule (Campbell et al. 1997, 300).

b. *The toy model*

But what about evidence for the emergence of life from such basic organic molecules? Although sophisticated computer algorithms show that this can happen, there is the more accessible toy model. Throw about ten thousand buttons on the floor. Randomly choose any two buttons and connect them with a piece of thread. Throw the connected button pair back on the floor. Repeat this with two other randomly chosen buttons. As you continue the game, at first the odds are very high that you will pick up two buttons that you had not picked up before. After a time, though, one of the buttons you randomly choose will be one that is tied to another with a piece of thread. So after you finish with this task in the game, you will have your first triplet of buttons connected with thread pieces. As you continue this marathon game, eventually you will more frequently choose a button connected to a larger cluster of interconnected buttons. Some buttons are not connected to other buttons. Others are twinned, others in triplets, others in clusters.

What has been found modeling this game with a computerized algorithm is that eventually, and very consistently, a phase transition occurs when the ratio of thread pieces to buttons passes 0.5 (10,000 buttons, 5,000 thread pieces). At that point, a giant cluster of buttons suddenly forms. Suddenly, most of the small, moderately sized, and larger clusters become interconnected into one giant structure. Most of the buttons are directly or indirectly connected. As the giant structure grows even larger, its rate of growth slows since the number of remaining single buttons and small clusters decreases. The rather sudden change in size of the largest button cluster after the thread-to-button ratio passes 0.5 mimics Kauffman's view of the phase transition of interconnected relationships among molecules in the "chemical soup" from which life first emerged. The buttons in that chemical amalgamation were molecules, and the threads were connections, such as bonds formed between and among molecules. The point is that when the ratio of connections among molecules (number and kinds of molecules) increases, a threshold ratio of connections to molecules is arrived at and surpassed. Once this (phase transition) occurs, there are suddenly so many interconnected molecules that a vast web emerges of diverse molecules interconnected by "threads." Once a large-enough number of chemical interconnections are catalyzed in a system composed of enough interconnected autocatalytic sets, a vast web of catalyzed chemical events suddenly emerges and maintains its robust, resistant-to-perturbation order. Such a web, Kauffman claims, is almost certainly autocatalytic, self-sustaining, and alive (Kaufmann, 1995, 54–58).

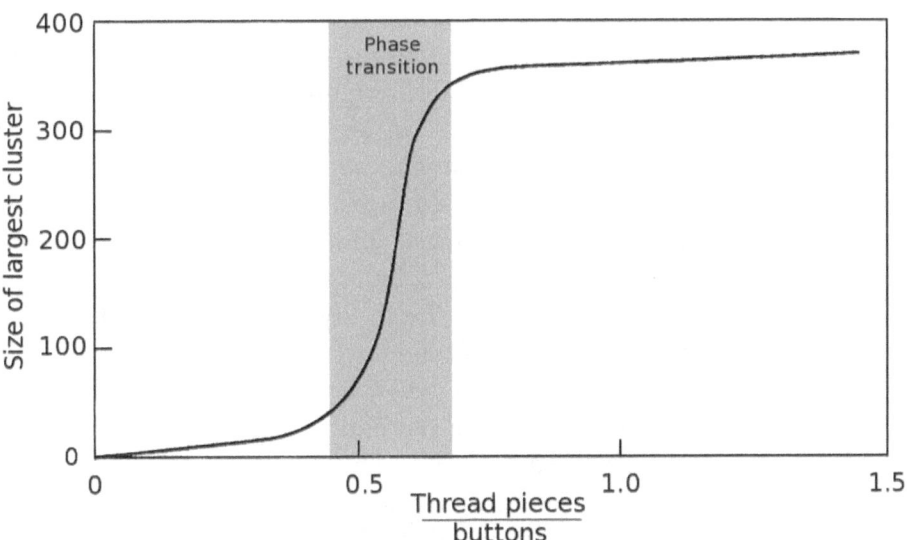

**Figure 6.4. A phase transition curve. To the left of the grayed phase transition area, each successive button-tying event does not increase the size of the largest cluster much on average. Suddenly, when the number of thread pieces used to connect buttons reaches half (0.5) of the number of buttons, the size of the largest cluster increases dramatically with relatively few subsequent button-tying events. To the right of the phase transition, the system is entering another stable state as the one to the left of the phase transition, in which adding successive button-tying events again does not much affect the size of the now-dominant cluster. In sufficiently complex, interrelating, nonequilibriium molecular systems, with molecules instead of buttons and connections instead of threads, after such phase transitions, new "order for free" emerges, governed by laws not reducible to those governing the system preceding the phase transition. Life may have emerged from a nonliving chemical system in this way (Adapted from Kauffman 1995, 57).**

*The improbability of life emerging simply by chance*

In fact, the chance that anything as complex as a *Mycoplasma* bacterial cell (one of the simplest living organisms) emerging by a process of contingent, random chance is diminishingly small. Robert Shapiro, in his book *Origins*, calculates that there may have been $10^{51}$ attempts by the universe to create life, given only random chance. That's one with fifty-one zeros behind it. A very large number. But wait until you hear what comes next. What about the probability for success in each of those trials? Shapiro attempted to calculate the odds of the emergence by chance of a creature at the level of simplicity of the bacterium *E. coli*. *Escherichia coli* is a pretty simple organism with about 4,400 to 5,400 genes, compared to the 485 that *M. genitalium* has (Kimball 2009). Humans have about 25,000 genes at latest count. Shapiro follows an argument by the astronomers Fred Hoyle and N. C. Wickramasinghe who started small, trying to calculate the chances of just getting a functioning enzyme. Enzymes are proteins, composed of chains of simpler molecules called amino acids. Enzymes are made and used as catalysts by organisms. Catalysts promote vital chemical reactions that sustain life in organisms' bodies. There are twenty kinds of amino acids. The first question is, what are the chances of obtaining an actual bacterial enzyme with two hundred amino acids? If the amino acids are randomly selected and arranged in random order, then the answer is computed by multiplying the probability for each correct amino acid in the sequence (1 chance in 20) two hundred times. The solution is 1 chance in $10^{200}$, or one with two hundred zeros behind it, an unimaginably small probability; even smaller than $10^{51}$ is large. But since some amino acids can replace one another in amino acid sequences, the authors concede to a probability of 1 in $10^{20}$. A problem, though, is that one functional enzyme is not nearly sufficient for a bacterium to replicate itself: a cardinal condition in the definition of a living organism. Instead, about two thousand functioning enzymes are necessary. The odds that this would happen by chance are 1 in $10^{20 \times 2000}$, or 1 in $10^{40,000}$ one with forty thousand zeros behind it). The denominator of this number is beyond what the astronomers use in their calculations of the universe. The most common atom in the universe is the hydrogen atom. There are only $10^{60}$ hydrogen atoms in the universe. Kauffman puts the conclusion succinctly: "If the total number of trials for life to get going is 'only' (quotes mine) $10^{51}$, and the chances are 1 in $10^{40,000}$, then life just could not have occurred." This seems a reasonable statement. It also seems that the powers of self-organization that we've looked at in the previous pages are tantalizing, compelling, even if not yet conclusive, by Kauffman's own reckoning. As he continues, there are reasons to believe that whenever a "soup" has enough chemicals consisting of enough different kinds of molecules, a living metabolism will crystallize from the "broth." If correct, this means that the first living organism, with all its intricate, impossibly complex, beautiful biochemical patterning wasn't built over improbable eons, one component at a time; but rather, it emerged fully formed, whole, crystallized out of the broth (Kauffman 1995, 43–45).

# Summary

☐ For Kauffman, the order in living systems doesn't just arise from random, contingent factors upon which natural selection works. Instead, natural selection works on order that has arisen from a source in the universe. In other words, complex living organisms are not tinkered together by chance but, rather, are to be expected to emerge in such a universe as ours, even if the precise details of any particular emergent pattern can never be known before the pattern emerges.

☐ The new science of complexity explains the relative suddenness in the fossil record in which whole new groups of organisms arise. One such example of the sudden rise of many diverse animal forms is the Cambrian explosion, between 550 and 480 million years ago. Before 580 million years ago, life had persisted as single-celled organisms for billions of years. About 580 million years ago, evidence for simple, soft-bodied multicellular organisms began to appear. Then, beginning around 550 million years ago until about 480 million years ago, the fossil record shows that the abundance and diversity of entirely new, more complex life-forms appeared so relatively suddenly that paleontologists call the period an explosion. According to Kauffman, such complexity can arise when relational entities group and a network of relationships develops. This process continues until a phase transition is reached, in which the number of relationships increases more quickly than the number of new entities introduced into the system. Explosively fast complexification occurs; and new "order for free," robust against perturbation, emerges. The fast complexification causes the system to lose some of its stability and approach becoming chaotic. It is at this edge of chaos that robust "order for free" emerges, which may express itself as spontaneous emergent novelty.

☐ Life itself likely emerged from prelife in a similar way to that described above. There is growing evidence that it emerged and spread quickly in a relatively short period.

☐ The autocatalytic network helps to explain how life may have emerged from nonlife. These complex molecular networks are composed of molecules, which in the presence of other molecules can catalyze or affect the synthesis of more molecules like themselves.

Thus, such a molecular network can propagate itself over time, given that a flow of "food" molecules is available to the network. Life is composed of open, nonequilibrium chemical networks composed of many autocatalytic sets.

☐ Our 256 different cell types that compose our bodies are emergent patterns of chemical reaction networks.

☐ Shapiro, following an argument by astronomers Hoyle and Wickramasinghe, found that the chance of the most simple bacterial organism emerging from a prebiotic soup is 1 in $10^{40,000}$. The latter number is beyond what astronomers use in their calculations of the universe. In other words, according to Kauffman, if left to chance alone, life could not have come about.

☐ Because of primal universal laws of complexity, as yet admittedly hypothetical, Kauffman says that there are good reasons to think that whenever a chemical soup is diverse and complex enough, living metabolism will crystallize from this broth relatively suddenly.

☐ A critical mass of complex interrelationships among molecular components, triggering a phase transition, led to the condensation of new order and qualitatively new kinds of organized wholes—the order robust, resilient, and emerging as though inexplicably, freely—as life did probably 3.8 billion years ago. That's the age of the oldest continental crust on earth today. The oldest evidence of life, a version of carbon atoms known as $C_{12}$, normally associated with the presence of life, has been found in Greenland rock of that age. The solar system, including the sun, earth, and other planets, is about 4.5 billion years old. Less than a billion years later, life had probably already emerged. A billion years later, it certainly had. Three-and-a-half-billion-year-old fossil bacteria have been found in Western Australia and South Africa, remarkably similar to modern oxygen-producing cyanobacteria (Christian 2004, 58, 109).

☐ Evidence strongly suggests that this was impossible by Darwinian laws alone. Deep laws of the universe, preceding life itself, almost certainly were and are at play. These laws work on complex, dynamically interrelational, open, nonequilibrium systems like living systems that matter and energy flow through. As these networked systems grow in size and complexity, they become less stable and become poised at the edge of the chaotic regime.

☐ In this narrowly defined ledge, there is the possibility of the emergence of creative novelty as networked interrelationships

JOSEPH FORTIER

complexify in them and order condenses out of them. These systems act similarly to computerized algorithms in that there are no predictive shortcuts as to what the new, emergent order (solution) will look like. In both cases, the process is its own best shortcut. One must simply watch this process through to its outcome.

☐ New laws emerge with new "order for free" that are not reducible to laws governing the ingredients of the new order. The evidence strongly suggests that deep laws are responsible for most of the order that has emerged within the history of life on Earth, and that natural selection works in tandem with these laws, tweaking the emergent order in specific directions.

# Glossary

**algorithm**. A method for solving a problem using a finite sequence of instructions, used in mathematics, computing, and linguistics.

**autocatalytic network**. A chemical system such as a living system, composed of many autocatalytic sets whose functioning is enmeshed with one another into an integrated network that depends on a constant supply of various chemical ingredients to remain operative.

**autocatalytic set**. A chemical system in which some molecules can catalyze their own formation.

**catalyst**. A molecule that facilitates the occurrence of a chemical reaction.

**chaos theory**. A theory that was developed from the observation of weather patterns that describes the infinite succession of patterns in certain (chaotic) systems in which no two patterns are identical and patterns are fleeting, constantly changing.

**gradualism**. The theory that evolution has occurred as a constant, gradual process of change.

**niche**. An opportunity within an ecosystem for a specific kind of organism to live within.

**nonequilibrium system**. A type of open system, such as a living system, in which the flux of matter and energy entering and leaving the system constantly changes.

**open system**. A system of interrelating components, such as a living system, that relies on matter and energy from outside itself to maintain its functioning and organization.

# References

Campbell N, J. B. Reece, and E. J. Simon. 2007. *Essential Biology Third Edition*. San Francisco, California USA: Pearson/Benjamin Cummings.

Christian D. 2004. *Maps of Time*. Berkeley, California: University of California Press.

Futuyma D. 2005. *Evolution*. Sunderland, Massachusetts, USA: Sinauer Associates.

Kauffman S. 1995. *At Home in the Universe; The Search for the Laws of Self-Organization and Complexity*. New York: Oxford University Press.

Kauffman S. 2008. *Reinventing the Sacred: A New View of Science, Reason, and Religion*. Philadelphia, Pennsylvania USA: Basic Books.

Kimball J. 2009. "Genome Sizes." Last modified February 3, 2011. *http://users.rcn.com/jkimball.ma.ultranet/BiologyPages/G/GenomeSizes.html*.

Medscape. 2004. "Mycoplasma infections." Last modified 2011. http://health.discovery.com/encyclopedias/illnesses.html?article=355

Microbewiki. 2011. "Mycoplasma genitalum." Last modified July 1, 2011. http://microbewiki.kenyon.edu/index.php/Mycoplasma_genitalium

National Museum of Natural History. 2009. "Pikaia." Last modified 2011. http://paleobiology.si.edu/burgess/pikaia.html

# CHAPTER 7

# Pierre Teilhard De Chardin, Evolution, Complexity, Consciousness, and Spirit

## Teilhard de Chardin: a brief biographical sketch

PIERRE TEILHARD DE Chardin was a French Jesuit paleontologist and geologist who lived from 1881 to 1955. He was an avid thinker and writer from his earliest years as a Catholic priest in the Jesuit religious order, especially concerning natural history and evolution. As a scientist, he is credited as being a codiscoverer of fossil Peking man, now known as *Homo habilis*, a species in the evolutionary lineage of modern humanity (*Homo sapiens*). In 1925, some of his fellow Jesuits requested from him a written essay on his thoughts concerning biological evolution and the book of Genesis concerning original sin. One of them later evidently submitted the essay to the Jesuit superior general in Rome. Teilhard was told, under his vow of obedience as a Jesuit, to leave his teaching post in Paris and sign a formal statement declaring that he would never again write or say anything contrary to an ahistorical literalist view of the Genesis creation narratives. He wrote an alternative version that he felt would not compromise his intellectual honesty and in his mind would not cause offense in Rome. The Jesuit superior general's response, however, was to send him immediately not only from his teaching post but also from France even after Teilhard signed not his drafted formal statement, but one even more rigidly antiscience than the first one with which he'd been presented. After consultation with a trusted friend and in considerable distress, Teilhard signed the statement but in the spirit of a physical action that would signal a gesture of fidelity rather than a symbol of intellectual assent (Lukas and Lukas 1977, 69–95). It was the Galileo affair all over again.

Teilhard remained a Jesuit and a research scientist. He remained faithful to his spiritual and intellectual integrity as well as could be expected in the face of adversity from the Jesuit and Roman Catholic authorities at the time. In later years, Teilhard's writings were prolific and passionate in his advocacy for an intellectually honest integration of science and Christian faith, although the Catholic and Jesuit authorities denied him the right to publication during his lifetime. His grand manifesto of his position on evolution and Christian faith, *The Phenomenon of Man*, was published in the same year of his death—1955.

Unfortunately, since Teilhard was not allowed to publish before his death, he was deprived of the opportunity to dialogue with the larger professional community in his fields of biology, theology, and philosophy with respect to his writing. If he had been able to, some of his assumptions, such as that of orthogenesis, or the evolution of one form directly from another with no branching pattern of descent, might have been challenged and modified. Another of these assumptions is that evolution is directional and goal oriented. Since there is no empirical evidence for this, it is important to say this. At times, Teilhard conflates empirical science with his faith, which had been informed by biological evolution and, thus, falls into these assumptions. In passages where this occurs, his positions would have been strengthened if he had made necessary distinctions between rigorous empirical science (which he participated in) and theological speculation that is informed by biological science. I will summarize and critique the major points of *The Phenomenon of Man* below, compare and contrast them with Kauffman's thought, and discuss their significance for a theology of nature that integrates biological evolution and Christian faith.

## The teapot on the stove

Before beginning discussion of Teilhard's position on evolution and spirituality, it may be helpful to note that there can be two different but non-conflicting, parallel explanations for the same phenomenon. To use an example that I heard Dr. John Haught present a few years ago, imagine a teapot on the stove, the water inside the pot boiling. The question is, why is the water boiling? If a given answer were "Because the water molecules have absorbed sufficient heat energy from the stove that they are flying apart at high velocities, and thus, the liquid state of the water is changing to the gaseous state," this would of course be correct. Another correct answer might be "Because I wanted tea." Both answers may be correct, yet neither knowable from the other. They are parallel explanations in that they don't intersect. In the same way, it is possible that there may be parallel explanations for the emergences of life, of social cooperation, of the faculty for learning, of the faculty for awareness of one's awareness, and for the faculty to sense the reality of the transcendent.

# The science-religion dilemma as seen by Teilhard

At the beginning of the second chapter of *The Phenomenon of Man*, Teilhard describes the challenge he confronts in the two conflicting adversaries of rigid scientific materialism and rigidly doctrinaire religionists. On the one hand, he sees that the materialists insist on describing objects as though their behavior only consisted of external actions in transient relationships. No deeper law, theme, or wholeness than random jostling and chance association. Of course, this does not entirely describe Kauffman's position on reductionism, except that Kauffman does not seem to see anything other than external relationships. On the other hand, he also perceives the advocates of a sort of rigid spiritual interpretation who stubbornly refuse to go beyond a view that only sees things as being shut in upon themselves in their immanent workings—the eternal, unchanging present. Neither listens to the other; both avoid dialogue, neither perceives the entire problem. So Teilhard's goal is to address both sides of the debate by describing the genesis of the universe in terms of its basic interrelated wholeness and in terms of an interior aspect of the whole and of each constituent corpuscle: the aspect of "within." He accomplishes this by focusing on the human lineage. He is faulted by many modern evolutionists for this anthropocentric focus on the orthogenesis of the human lineage, probably justifiably. In Teilhard's defense, his specialization in paleontology was human evolution. He focused on what he knew best and focused to make his point as clearly as possible: that the universe is a spiritually inspired creation that expresses its own spiritual inspiration in its relationship with *Omega* through the process of the evolution of consciousness, which reaches its apex in the human species.

## The "within" and the unity of spirit and matter: Teilhard's monism

Teilhard develops the notion that all entities in the universe have an interior reserve, an inwardness that he calls the "within." The "within" is a sort of principle of subjective relationship, a primordium of consciousness and of spirit. He derives this idea as follows.

The first point Teilhard makes is that the "cosmic corpuscles," or atoms and their component particles, are only discernible by virtue of their effects on other such particles. Thus, the universe is deeply woven

JOSEPH FORTIER

together and relational at its most primordial level. Teilhard writes that each element of the universe is "woven from all the others." One aspect of this "wovenness" is that each element is composed of subunits (molecules of atoms, atoms of subatomic particles, subatomic particles of quarks), and thus at this primordial level of matter, each atom is "woven" from the same kinds of particles as other atoms. The second aspect of the "wovenness" of the universe is the influence of "unities of higher order" that incorporate the element into their own ends as living cells incorporate molecules. The universe is thus a system, a whole, of a piece (Teilhard de Chardin 1959, 41–44). It can be seen how this notion of a dynamically interrelational system anticipates Kauffman's thought (chapter 6).

Secondly, it is possible to infer general qualities of matter from qualities that slowly emerge as one moves from a given level of some quality of matter to another. For Teilhard, this idea is distinct from complexification, which is very similar to Kauffman's "complexity," and will be discussed in detail below. Teilhard notes that among chemical elements considered stable, stability and longevity *appear* to be the rule but that the idea of absolute stability and longevity has been forever altered by the discovery of heavy, radioactive substances such as radium, which decay over time (chapter 4, radiometric dating). What would have happened to physics, he asks, if radium had been considered abnormal? Then we would never have come to see that other elements, whose levels of radioactive emissions are not so easily measured as radium, also share with radium the property of radioactive decay although at such minute levels as to be virtually imperceptible. In the same way, in normal human experience, velocity doesn't change the nature of matter. However, at the extremely high velocities at which atomic particles move, their masses are profoundly changed. Thus, we come to know that mass and velocity are related, one affecting the other, even though at imperceptibly tiny levels in ordinary experience, but at increasing levels as a particle's amount of mass under observation increases as velocity increases. Also along these lines, the mountains and stars symbolize stability, changelessness when observed within the timescale of a human lifetime; but over a greater scale of time, the earth's crust constantly changes, while "the heavens sweep us along in a cyclone of stars" (Teilhard de Chardin 1959, 54–55).

What if we apply these two themes of (1) the woven, interrelated unity of the universe and (2) the basic similarity of behavior of the stuff of the universe—to consciousness? We may be led to the possibility that consciousness itself, like radioactivity, like the effect of velocity on mass in atomic particles, is not completely absent in the simplest of entities, even though we associate it with only the most complex of mammals. In other words, the apparent, observable presence of some subjective capacity for relationship at a given level of complexity in organization of nature disappears into obscurity as one passes from that level, or scale, through successively less complex levels (scales), to a level or scale distant from that at which the quality can be observed. This is what Teilhard is getting at when he describes the panpsychism, or presence of consciousness, albeit at various levels, in the universe. This is what he means by the "within" of every entity, from the tiniest subatomic particle to the largest whale (Teilhard de Chardin 1959, 55–56). According to Teilhard, the subjective interiority of the "within" is a sort of protoconsciousness at simple levels of organization and exists everywhere. In very simple particles, the "within" is so small that we can't perceive it. As the complexity of exterior, objectively observable interrelationships that compose an entity increases (e.g., between and among particles such as atoms and molecules), subjective consciousness becomes increasingly apparent (Teilhard de Chardin 1970, 156). This coincidence of the rise of consciousness with the rise of complexity is "complexity-consciousness."

We see that with respect to matter, for Teilhard, the universe is a relational, interwoven monad, composed of diverse but related, interactive particles, all of which have at least the subliminal glow of protoconsciousness. The universe is a system in which its components are in dynamic relationship since the atomic particles of which it consists are composed of interactive subatomic particles and are influenced, related with, dominated by more complex aggregate particles, such as complex molecules, suns, planets, and living entities.

## The activity of the "within": energy and how complexification happens

Teilhard describes the energy in the universe as psychic in nature. By this, he seems to mean relational in a way not completely quantifiable. The energy in a particular entity is divided into two components: tangential energy, which links entities such as atoms and molecules with others of the same order of complexity and centricity, and radial energy, which draws the entity toward increasing complexity and centricity. By centricity, Teilhard means the degree of capacity for conscious relationship. Centricity increases with greater complexity and condensation of order in living systems (by which he means

JOSEPH FORTIER

something closely similar to what occurs after a phase transition as described in chapter 6). Teilhard notes that the tendency for a direct positive relationship between complexity and centricity in the evolution of life is observable. As the organisms in an evolutionary lineage become more complex in matter and internal energetic relationships over time, their behavior becomes less easy to quantify, more descriptive in nature. This, for Teilhard, is evidence of movement toward consciousness. Even in plants, he writes, this tendency toward greater centricity is observed (Teilhard 1959, 55).

It is the tangential component of energy that is accessible to empirical science. The radial component is more difficult to define, except to say that it is spiritual in nature and capable of reading information not accessible to empirical observation until after the fact of an increase in complexity, centricity, and emergence of novelty based on a newly condensed degree of order.

Teilhard's description of the process of complexification is similar to Kauffman's description of the emergence of qualitatively novel, more complex systems with new "order for free" when open, dynamically interrelating networks composed of autocatalytic sets undergo a phase transition of complexity. A chemical autocatalytic set is a complex network composed of many kinds of chemicals, some of which can catalyze the formation of others (chapter 6). According to both Kauffman and Teilhard, the shape that the new emergent order takes can never be predicted a priori, or before the order actually emerges. For Kauffman, the principal cause is deep law within the universe. For Teilhard, it is the interplay of radial and tangential energy, as I will explain.

Once a given threshold of complexity is achieved at some point in the evolutionary trajectory of entities at a given level of the complexity of an autocatalytic network (a prebiotic "soup," a group of unicellular organisms, the nervous system of a group of animals), an entity in this initial state of complexity, with some free tangential energy (the component accessible to empirical observation), will be able to increase its internal complexity in association with neighboring entities. One can imagine a large cluster of organic molecules in a prebiotic soup incorporating other organic molecules into the cluster, and thus increasing its own complexity. It is a universally accepted theory that more modern, complex nonbacterial cells such as those in our bodies have descended from simpler bacterial cells, in which some of these cells ingested those of another kind and, thus, increased their internal complexity (Futuyma 2005, 95). Teilhard observes that there is a direct positive relationship between complexity and centricity as noted above. Radial energy (a sort of spiritual energy, not directly measurable; responsible for the rise of consciousness) is activated when some of the tangential energy in the system becomes free and augments radial energy. The radial energy is responsible for forming a "new arrangement in the tangential field" (Teilhard 1959, 65).

## Teilhard and Kauffman

At this point, one might ask, "Well, who's correct? Teilhard or Kauffman?" Or one might even feel the impulse to take a position. In fact, they may both be correct. Perhaps a deep law such as what Kauffman predicts will be discovered that explains the mysterious self-ordering power in our universe. By definition, the details of the final outcome of any original, specific self-ordering event of complex systems such as life can't be known a priori before the new order condenses. These events occur in unstable, near-chaotic conditions. The order that does arise arises spontaneously, albeit this order becomes robust, resistant to perturbation after the fact (Kauffman 1995, 8, 18, 75–80).

Such a law would not necessarily conflict with Teilhard's proposal concerning "within" and radial energy. By definition, such laws are very general and allow for great creativity and spontaneity (Kauffman 1995, 15). Teilhard's concepts carry with them a deep relational intimacy and would account for not only general laws of the emergence of spontaneous, robust order but also might account to a greater extent for details within each event of emergent novel order. In Teilhard's mind, this process would never preclude what is observable by the scientific method such as natural selection as this would violate the integrity of science.

As with the two parallel answers to the question why is the water boiling, neither proposal negates the other. In the minds of both, each proposes an explanation for the complexification of the interrelationships of matter and energy in dynamically interrelating systems such as living systems. Kauffman's proposal is based on empirical evidence only. It admits to an as yet undiscovered deep law of the universe. Teilhard's proposal is based on empirical evidence and spiritual sensibility. The fact that Kauffman's proposal is based only on empirical evidence gives it valuable academic credibility in the established scientific community. The fact that Teilhard's proposal incorporates spiritual sensibility weakens its credibility in the established scientific community, an effect that I don't think Teilhard ever quite understood. However, it provides a brilliant integration of science and spirituality at a threshold at which the two areas of investigation meet. It invites scientific inquirers to consider the possibility of a spiritual side of nature without compromising their work as defined by rigorous empirical inquiry and to religionists to allow their worldviews to be informed by scientific inquiry and discovery of nature. Teilhard's work anticipated emergent novelty within complex, dynamic systems such as life, independently of natural selection, fifty years before Kauffman's work, without the aid of modern computation.

## Teilhard's and Kauffman's insights compared

Teilhard anticipated the following four qualities in Kauffman's description of the emergent order of novel complex living systems in the evolutionary process. First, in a complex, dynamically interrelating system at the edge of chaos, large networks spontaneously condense into one or more ordered patterns. Because of the huge number of possible specific outcomes of such patterns and the relatively sudden spontaneity of this condensation of new order, the actual outcome of what particular ordered pattern emerges is impossible to predict. It is similar to writing and executing a mathematical computer algorithm for a complex nonequilibrium system at the edge of chaos on a computer. The process of the computer in finding a solution is its own shortest solution (chapter 6). Second, once this unpredictable, spontaneous new order has condensed, it maintains its robust integrity. Third, in spite of the robust integrity of this newly emergent order, it is flexible. Fourth, this newly condensed emergent order only emerges after the amount of complexity undergoes a phase transition of relationships between entities in the system (e.g., number and kinds of molecules), during which the *number of relationships* between entities in the system increases rapidly with respect to the *number of entities* in the system.

Teilhard wrote an interesting passage in *The Phenomenon of Man* in which, describing the first living cells that emerged from nonliving matter, he anticipated the first three of the above points. Here is the passage: "Next is fixity . . . indefinite as are the possible modulations of the fundamental theme, inexhaustible as are the various forms it assumes in nature, the cell remains in all cases essentially true to itself" (Teilhard 1959, 87). In the first clause of the second sentence in this passage, beginning with *indefinite*, he anticipates Kauffman's first quality of emergent patterns in complex nonequilibrium systems at the edge of chaos (above paragraph), such as complex chemical systems and living systems. The "modulations," which are the possible outcomes for the potential "fundamental theme" of the new emergent order, in this case whatever will emerge from the primordial complex chemical soup, are "indefinite." "Indefinite" refers to the a priori inability to predict the final outcome and the tremendous number of possibilities for the final outcome of the event of emergent new order.

For Teilhard, Kauffman's second quality of emergent patterns in these complex systems, which is the robust yet flexible integrity of the newly emergent order, is "fixity" (Teilhard 1959, 87). He describes the cell as remaining "in all cases essentially true to itself." He also qualifies this fixity as having flexibility, thus anticipating Kauffman's third quality of these complex systems. For Teilhard, this flexibility is expressed as an "inexhaustible" variation of cell forms in nature.

Teilhard's "thresholds of complexification," anticipating Kauffman's fourth quality of emergent patterns in complex systems, are those stages in the process of biological evolution at which the interrelationships of particles in a system growing in complexity cross some critical threshold of intensity and the conglomeration becomes a more highly ordered unit. By "critical thresholds," Teilhard's meaning is the same as that used in the natural sciences (Teilhard de Chardin 1970, 33–34, 213, 284). Teilhard describes the threshold of complexity over which human reflective thought emerges as a "human critical point" (Teilhard de Chardin 1959, 88). For Kauffman, this fourth quality of the process of evolution of complex systems is condensation of new "order for free," after the process has passed through a phase transition with respect to the ratio of the number of interrelationships to the number of entities in the system (chapter 6). By "human critical point," Teilhard evidently intends to convey the metaphor of a physical change of state to the transition from nonhuman consciousness to human reflective consciousness. Thus, "heat of vaporization" is a more precise term for the metaphor than "critical point." The *heat of vaporization* of a substance is the *additional amount of energy* needed to change the physical state of the substance from liquid to gas once the substance has reached a temperature at which this phase transition may occur. Thus, for water, the additional energy needed for liquid water to turn into water vapor after the temperature nears 100°C is the heat of vaporization. There is a close relationship between the critical point (Teilhard's usage) and the phase transition (Kauffman's usage): both refer to what occurs at a threshold of change in physical state of a substance. Both Kauffman and Teilhard use the metaphor of change in physical state in reference to emergent, new complex order in living systems.

# Thresholds of complexification, consciousness, and the role of the Omega

## Sudden appearance of new order in living systems and the Cambrian explosion

Teilhard couldn't have known about the Cambrian explosion, or dramatic increase in diversity of animal life during the Cambrian period about 542–530 million years ago (chapter 6) since the scientific community didn't understand fossil finds from this period until about 1970. Around that time, H. W. Whittington published work on his reinvestigation of Cambrian fossils found by earlier exploration from 1909 and later, and by his own exploration of the Burgess shale deposits in the Canadian Yoho National Park near Field, British Columbia (Gould 1989). As we saw in chapter 6, Kauffman used this evidence for an explosive, relatively sudden emergence of new complex order and design in animal life to support these themes in the science of complexity.

But Teilhard was aware of evidence for the evolution of life from nonliving chemical aggregates. At the time of finishing *The Phenomenon of Man*, before his death, he may have been aware of the Miller-Urey experiment. In their famous experiment, these two scientists demonstrated that organic molecules, which many had thought could only be produced by living things, could be spontaneously produced under conditions similar to those of the early earth. Using this idea of the rise of the first bacterial life from nonliving matter, Teilhard speculated on the probable sudden appearance of life on earth, complexification, and emergent new order (Teilhard de Chardin 1959, 77–97).

Teilhard saw that these newly emergent cells were characterized by their complex, diverse chemical composition and the networks of relationships among a cell's chemical compounds: "forces of viscosity, osmosis, and catalysis which characterize matter when molecular groupings have reached an advanced stage" (Teilhard 1959, 87). He describes the cell as a *complex*, a word for which synonyms are *system* and *network*—Kauffman's usage.

The notion of the relatively sudden appearances of new forms of life in the fossil record wasn't commonly accepted in scientific circles until after Stephen Jay Gould and Niles Eldredge published it in the 1970s, when they coined the term "punctuated equilibrium." Nonetheless, Teilhard pioneered the concept (Morowitz 1997, 26) and understood the relative suddenness by which major events of emergent new order with novel design may occur. Using his idea of the psychic nature of matter, that is, that even the simplest particles of matter possess an interior, preconscious, relational "within," Teilhard develops the speculation that the evolution of emergent, newly complex order developed in "jumps." For instance, he notes that the emergence of life from nonlife, in which the organized whole governs the behavior of its constituent molecular parts, shows a qualitative jump in "awakening," a "sudden rise" of "the degree of interiority" to another level (Teilhard 1959, 89). These jumps of degree of interiority are, for Teilhard, strongly correlated with relatively sudden jumps of complexification of the "without" of things in evolutionary history. Teilhard in effect anticipated punctuated equilibrium (Gould 2007; Eldredge and Gould 1972; Gould and Eldredge 1977) "when Gould and Eldredge were in short pants" (Morowitz 1997, 26).

A further aspect of the sudden appearance theme in Teilhard's speculation was his notion that "life no sooner started than it swarmed" since the early earth must have been in a "state of biological super-tension" (Teilhard 1959, 92–93). These ideas also anticipate Kauffman's use by example of the Cambrian explosion of new forms of animal life (unknown in Teilhard's time, as previously noted) to describe the sudden rise of complexity in animal body design and consequent rapid proliferation of these animal forms (chapter 6).

## Complexity-consciousness: exterior emergence of complex order and interior emergence of consciousness

Teilhard suggests the emergence of various spheres of new emergent order on the earth. For Teilhard, there is always a correlation of increasing complexity with increasing consciousness in biological evolution, which he describes as a law (Teilhard de Chardin 1959, 287, 300–302; 1970 71–72, 144, 155, 324; also see 'The 'within' and the unity of spirit and matter: Teilhard's monism"), much as Kauffman describes "deep, hidden law" (chapter 6). The "within" moves farther along toward becoming more fully conscious with each evolutionary event in which complexification occurs, especially with respect to complexification of the vertebrate nervous system (Teilhard de Chardin 1959, 144).

For Teilhard, these complexification events involve small or large jumps in the fashion of Per Bak's self-organized criticality (see Per Bak, chapter 6). Large jumps over major thresholds of complexity and "within"

JOSEPH FORTIER

are involved in crossing into the biosphere (Teilhard de Chardin 1959, 78–79, 112; 1970, 64–65, 85) and the noosphere (Teilhard de Chardin 1959, 180–184). We explored the jump into the biosphere above, where the emergence of complex living systems from complex molecular systems was discussed here and in chapter 6. Let us now turn our attention to the noosphere.

In the long natural history of evolutionary developments within the biosphere, including jumps of complexification accompanied by progressive development of "within," consciousness arose as lineages of organisms evolved through time. Psychogenesis, Teilhard's term for this biological process of the rise of consciousness, refers to these jumps. Psychogenesis involves steps in the complexification of nervous systems of animals (Teilhard de Chardin 1959, 142–148) and in the corresponding progressive development of intelligence (capacity to learn), particularly in the chordate lineage (Teilhard de Chardin 1959, 144). Evidently, Teilhard wasn't aware of nonvertebrates that are highly intelligent, such as the octopus (Scigliano 2003).

The noosphere emerged as a giant evolutionary jump within psychogenesis. Accompanying the complexification of vertebrate nervous systems was an increase in "within," or rise in consciousness (Teilhard de Chardin 1959, 147–160). Teilhard describes the way this process led to human self-reflective intelligence as an example of what he interprets as directed evolution, or orthogenesis. Within the mammals (vertebrate animals with hair and in which females have mammary or milk glands), the primates (lemurs, tarsiers, monkeys, hominids including the human lineage) arose about 50 million years ago. Interestingly, the most recent find of oldest New World primate fossil is named after Teilhard: *Teilhardina magnoliana* (Roach 2008). Unlike many other mammalian as well as nonmammalian lineages, the primate lineage remained relatively morphologically unspecialized. The primate lineage used the evolutionary freedom that this lack of morphological specialization gave it to "lift themselves through successive upthrusts to the very frontiers of intelligence" (Teilhard 1959, 159). This intelligence was gained by the process of cerebralization. The culmination of this process, for Teilhard, was the birth of thought in the human lineage (Teilhard 1959, 158–159). With the emergence of the human power to think as humans do came the emergence of the noosphere: uniquely human reflective consciousness capable of being articulated with highly developed language within a social context (Teilhard de Chardin 1959, 107, 180–182, 203–212).

For Teilhard, reflection is the power acquired by consciousness to go within the self, possess oneself as an object. One doesn't just know, but also, one knows oneself; one knows that one knows. One becomes conscious of one's own organization; one becomes clearly conscious as an individual (Teilhard 1959, 165). The noosphere grew with human socialization and communication between diverse human societies and cultures to become a spherical layer over that of the biosphere from which it emerged. This hominization process is consistent with the theme of biological complexification with accompanying development of "within" consciousness (Teilhard 1959, 182), a deep theme of development of the universe in Teilhard's thought, as we have seen.

*Hominization and culture: the noosphere*

The noosphere arose from a series of evolutionary jumps or steps from within the biosphere. Psychogenesis, including the rise of the ability of living organisms to learn with a sufficiently high level of outward complexification of the nervous system, as we have seen, is an observable manifestation of the "within" in Teilhard's thought. At some point in the psychogenesis of the human lineage, a critical point of complexification and interiority was achieved, and psychogenesis was led over a critical threshold of complexity. Noosphere emerged (Teilhard de Chardin 1959, 159–160).

With the emergence of noosphere or "thinking layer," the psychical aspect of the organism (*Homo sapiens*) becomes clearly defined as reflective consciousness. The effects of reflective consciousness on human behavior are recognizable and observable. Reflective consciousness becomes a principal part of the (human) organism (Teilhard de Chardin 1959, 176).

Manifestations of the noosphere, or layer of reflective consciousness over the earth, are the swelter of human cultures and constructs over the earth's surface in a diversity of languages, ethnicities, and political organizations (nations, tribes, etc). Each person is a facet, a sequin, a spark of reflection, each sending light reflected at a slightly different angle to the others. Thus, the whole of humanity has tremendous potential for reflective wisdom (Teilhard de Chardin 1959, 177–181). Teilhard sees this potential for the whole of humanity as a giant reflectively conscious network, as something we're moving toward. He sees us progressing toward a holistic awareness of the universe and of our relationship with it as though we are moving toward a point of convergence on which beams of consciousness emanating from all of us reflectively conscious humans are focused (Teilhard de Chardin 1959,

257–258). This notion of Teilhard's comes close to being deterministic, in the sense of a law of the universe and, thus, is reminiscent of Kauffman's notion of deep laws of the universe. For Teilhard, it is the yearning of the "within", which in human persons has become reflectively conscious as it is attracted, fascinated by a future wholeness at which we haven't yet arrived. Kauffman expresses a similar yearning when he expresses his hope for a "global spiritual space" in which honest, open dialogue between faiths and civilizations concerning the meaning of the sacred and the universe can take place (Kauffman 2008, 282). For Teilhard, this yearning will be answered in the fullness of time at the Omega point toward which the universe is attracted, headed. For Kauffman, it will be fulfilled in a secular enterprise in which religion will become viewed as anachronistic, an illusion—finally simply a vestigial appendage of the secular mind.

Teilhard contrasts human social evolution in the noosphere with biological evolution by noting that there is a pronounced tendency in social evolution for acquired characters in a LaMarckian sense to be passed on to the next generation, in contrast to biological evolution. Unlike beavers and insects with their stereotypical instinctual behavior that is inherited, much human heritage is cultural and is inherited with the aid of education. Evidence of this learned human behavior is in the form of constructions of materials for practical or beautifying purposes (homes, art) or written systems of thought (philosophy, science) or action (scientific discovery, dance, theater, sports, war, peacemaking, legislation). These constructs, in turn, augment human consciousness. The possibility exists to discover that our reflective consciousness, with all consciousness in the universe, is "the substance and heart of the evolutionary process" (Teilhard de Chardin 1959, 179). An example of how reflective human consciousness may be augmented by our actions, our creativity, is the action/reflection dynamic expressed in liberation theology. Liberation theology was invented by Catholic scholars working with the poor in Latin America and adapted elsewhere (McGrath 2006, 90; St. John 2008). The action/reflection dynamic consists of refining effectiveness in addressing social justice/human dignity issues. A first step is becoming involved in the lives of victims of injustice (action). The second step, which happens most effectively in conversation, involves becoming aware of how various economic and other human activities affect victims of oppression and injustice (reflection). This reflection process leads to more effective action on behalf of justice for the human dignity of victims, and the spiral becomes an evolving process of social evolution. Thus, Teilhard's recipe for augmentation of reflective human consciousness is presently used as a means for becoming more aware of the moral implications of human activity in order to remedy moral problems.

Teilhard uses the metaphor of the reproductive parts of the flower for the emergence of human soul. It is known that these parts—the pistil, stamens, sepals, and petals—are all modified leaves (Campbell et al. 2007, 330). Teilhard reflects that they are something very different from leaves but that they would have become leaves had they not been formed (embryologically in the case of the flower parts) under new influences, which changed their destiny. In a similar way, the human inflorescence has been transformed by the emergence of human soul within the emergence of noosphere from the process of psychogenesis. The animal and social primate qualities have undergone a metamorphosis with the crossing of a critical threshold of neurological complexification and consequent crossing of the "within" into the noosphere. These qualities remain in many respects the same but have been transformed, recast by the power of reflective consciousness including moral awareness (Teilhard de Chardin 1959, 179–180) although through the process of evolution (Teilhard, 184). With the human inflorescence of human mind and soul over the earth, the earth has grown a new skin and has found its soul (Teilhard, 183), a soul that expresses the yearning for fullness, completion of the multitude of "withins" that constitute the universe. The human soul is the voice for the prereflective primordial soul of the universe and is priestly in that it mediates and communicates in reflective conscious form, the yearning for full soulful expression of all existence.

Teilhard in an indirect way affirms the polygenic origin of the human species. It must be remembered that even as recently as 1947, the Roman Catholic hierarchy held out for monogenism: that all humans are descended from only two people, one of whom was the biblical Adam (Pope Pius XII 1947, No. 37). As we saw in chapter 2, almost all species, certainly all vertebrate species, have descended from a common *ancestral population*. In order for a population to grow in a genetically healthy way, it must have a critical mass of genetic diversity, which is only found in a sufficiently large population of individuals. Pope John Paul II affirmed this with his affirmation of the findings of biological science and evolution (John Paul II 1996). I suspect that Teilhard was vague and indirect about this for political reasons. Teilhard was bold enough to write that "the 'first man' is, and can only be, a *crowd* . . ." but then continues, "and his infancy is made up of thousands and thousands of years (Teilhard de Chardin 1959, 186). In his footnote to this sentence, he writes, "That is why the problem of monogenism in the strict sense of the word . . . seems to elude science as such by its very nature." Sadly, he felt that it was necessary to write about this in such a way that there could be more than one interpretation. The threshold into the noosphere, he writes, had to be crossed "in a single stride" (Teilhard, 186). In Kauffman's sense, the human species, and thus the noosphere of Teilhard, emerged whole.

To Teilhard, it is important to maintain a mindfulness of our own immersion in the evolutionary process. He doesn't make a rigid distinction between biological and social evolution. There is a tendency in the natural sciences, he writes, for scientists to presume that they're outside of the process of evolution, just watching. There can be an overly neat separation of subject and object in the shelter of the observatory, as though the scientist is not an element of the whole. Thus, it is important to be mindful of the fact that scientific intelligence itself is a product of the evolutionary process. Since Teilhard saw the evolutionary process as affecting both the "without" and "within," in order to maintain a coherent view of cosmogenesis including biological evolution, he felt that we need to be aware of the effects of evolution on our own soulfulness. Thus, he writes that the "fibers of cosmogenesis demand their prolongation in us such that it goes deeper than just the material." Here Teilhard moves into a controversial area. He speaks to the whole person of the scientist and requests that in incorporating the empirical method for scientific investigation (which is the only way to do scientific investigation), the scientist not deify that method but recognize that (s)he, as a person who is a scientist, is a larger reality, with a spiritual "within" than simply one composed of the "without" of things and only capable of relating with the "without" of all else. For Teilhard, as for John Haught, as we shall see later, our understanding of the evolutionary process is really only whole when we take into account our propensity for religious faith, the whole of our interiority. He quotes his friend and colleague Julian Huxley: "Man discovers that he is nothing else than evolution become conscious of itself" (Teilhard de Chardin 1959, 220–221).

Teilhard sees branching ramifications of the evolutionary process in a myriad of social phenomena such as the formation and spread of languages, growth and spread of new industries, and formulation and spread of philosophical and religious systems. These phenomena, he observes, are lifelike in their processes of formation and emergence and in their ability to reproduce themselves. When we reflect on how immersed our thinking processes and the processes of our cultural development are in the evolutionary process, we realize the universal, seamless continuity of biological evolution up to and including the threshold over into the noosphere (Teilhard de Chardin 1959, 223). It is noteworthy that Teilhard's notion of biological evolution is far more focused on the generalized, overarching Kauffmanian theme of emergence of complex systems than on the chance contingencies of Darwinian natural selection. While not excluding the role of Darwinian natural selection, he sees it as intertwined with the process of emergence of new "order for free," working out the details of that new order, as does Kauffman.

The noosphere, with its diverse modes of expression, involves systems that coalesce upon themselves in a way that is similar to the coalescence of interactive chemical networks into a living cell and similar to the interactions that occur among various kinds of organisms in an ecosystem. Unlike biological ecosystems, the emergence of reflective consciousness provides a new power to coalesce: the communication of thought. Teilhard shows his optimism when he describes his hope for the human community in the noosphere—an optimism that has been both questioned as naive and admired. Cultures in their various aspects and languages can unite before they speciate from each other. The divergences that persist are valuable as they contribute to the robust whole of human consciousness. Ethnic groups, peoples, and nations, when they consolidate, complete one another in a mutual fecundation (Teilhard de Chardin 1959, 241–242). Here Teilhard fails to make a distinction and weakens his case, something that might have been corrected had he been allowed to publish before his death. He postulates that the branching over time of the various human races, cultures, and languages has a *goal*: agglomeration and convergence and, thus, the solidarity of the human spirit (Teilhard, 243). Teilhard might have made the distinction between the similarity of the branching and differentiation of biological evolution and that of human cultural evolution on one hand and the essential difference in the reflectively conscious noosphere in that people become empowered by becoming increasingly capable of reflective self-determination (on the other hand). In the latter case, groups of people may become increasingly capable of deciding to make peace with each other and learn from each other—or not to do so. In any case, the notion of having a goal with respect to evolution is a faith statement, not a scientific statement. To the extent that we choose to actualize our potential to become increasingly conscious, aware, reflective on our experience and history, we may well become, as a species, more

JOSEPH FORTIER

in union with each other, more at peace, and more dynamically creative. Teilhard's position that this is a goal of evolution seems not only unscientific but also seems deterministic in a way that would compromise the freedom of reflective consciousness.

Teilhard's speculation that we humans may be headed toward greater unity with greater consciousness is drawn as a logical outcome from the general evolutionary pattern of the emergence of "complexification-consciousness" in natural history. The emergence of the unity of the cell evolved from chemical networks, and the emergence of the unity of the multicellular organism evolved from cellular networks. Thus, it seems to follow logically that with intensification of relationships among various peoples, a similar sort of unity may emerge. But Teilhard also uses in his sequence of emergences that of the ecosystem, with its symbioses among organisms from networks of organisms and abiotic environmental factors, toward a more profound unity of peoples. Thus his assertion that the thinking layer is headed toward a "more fundamental unity," a "single tissue," in contrast to the looser unity of an ecosystem is a faith speculation based on certain biological precedents rather than a rigorously argued logical outcome of cultural evolution (Teilhard de Chardin 1959, 242–243).

An obstacle to the unity of people that Teilhard sees is the tendency in modern societies to become depersonalized to the extent that individualism and the empirical materialist outlook dominate their cultures. In these societies, we tend to analyze and dissect reality. In this view, energy becomes dissociated from relationship and becomes a sort of new god in an impersonal universe. An effect of this mindset on us is a loss of respect for the person and his/her nature. Personality becomes a prison as the quality and importance of relationships with others is diminished. It becomes confused with individuality, cut off from authentic relationship with other personalities (Teilhard de Chardin 1959, 263). Genuine depth in relationship with others is vital for authentic personality to flower in the person so that life is truly dynamic and alive. Teilhard argues that the logic of evolution leads in the opposite direction from this sort of deadening, impersonal individualism (Teilhard, 257–259). As the evolutionary process produces what is whole (notice the agreement with Kauffman), so does the evolutionary process tend to produce whole human personalities, which find their completion in dynamic networks of relationship with other whole personalities.

## Toward Omega: noogenesis

Teilhard describes a threefold centering that he proposes is a property of every consciousness: (1) centering everything partially on the self, (2) being able to center self on self constantly, and (3) becoming more in association with other such centers. He proposes that this threefold theme of consciousness is in the process of becoming, of evolving in the noosphere; that it is an outcome of our living in a universe predicated on relationship among "centers" of "within" and that by the complex, concentrated order of our own conscious, reflective beings, we gather up this relational experience of the universe in conscious, reflective form. Thus, for Teilhard, we are undergoing a process of becoming a sort of spiritual network, with a holistic awareness of the universe and of humanity's relationship to it and to ourselves. This process of becoming is taking us toward a higher-ordered aggregation, toward a place of conscious convergence in the noosphere. Interestingly, this yearning for universal consensus echoes Kauffman's advocacy for dialogue among the whole spectrum of human beliefs and worldviews (Teilhard de Chardin 1959, 259; Kauffman 2008, 282). Teilhard envisions the noosphere, this layer over the earth of human reflective thought and conscious soul, as an open system in which matter and energy flow through (consistent with physics), but with a center. This center is for Teilhard the place of conscious convergence that transcends the noosphere, the Omega point, "beyond our souls" (Teilhard, 259–260).

It is when our focus becomes on this Omega point that we regain our subjectivity, our authentic, whole personality. As our focus becomes future oriented toward this Omega point and look beyond our souls, we begin a journey toward greater personalization, greater self-awareness. As humanity (the "grains of the noosphere") undertakes this project, the radial (spiritual) energy within it increases and the network of human reflective consciousness intensifies. The grains become perfected, fulfilled within this network with this quality of relationship (as opposed to the sparseness of a worldview only of analyzed, dissected, radical individualism). The mystery and depth of each grain of consciousness becomes more apparent. Of course, the Omega point is the effect of God on the universe, attracting us in our "within" by love, from the future, or at the nexus of the present and the future (Teilhard de Chardin 1959, 260–264). Teilhard also describes Omega as already in existence and operating "at the very core" of collective human thought (Teilhard de Chardin 1959, 291). So Omega is distant and transcendent, located in the future, but also immanent as well. As we will see in a subsequent chapter, John Haught's notion of God's activity incorporates these senses of the Teilhardian Omega as the attractive, noninterfering activity of God on the subjective "within" of the particle (atomic, organic, human). Omega presents itself in relationship with the universe through time. It is immanent in the sense of being actively present at each moment, to each particle, at the nexus of the present and the future, and transcendent in the sense of being final, in the future, beyond time, the completion and fulfillment of all existence in its relationship with God, the ultimate source of being, relationality, and love.

*Love, reflective consciousness, and Omega*

Like radioactivity and consciousness, Teilhard sees love as another aspect of things that has its primordia in simpler aggregations of matter but becomes a more dominant aspect of a thing as complexity increases. Like consciousness, love is a quality of the "within." Love is evident in mammals, which care for their young (I wonder if he knew that male giant water bugs also care for their young), feel sexual passion, and in some, feel social solidarity. Further down the biological continuum of complex organization, love becomes more obscure. The "within" of simpler aggregations of matter, such as atoms, molecules, and even bacteria, is the primordium of love since the "within," an "internal propensity to unite," makes love possible higher up in the continuum of complexity. Teilhard defines love as "a trace marked on the heart of an element by the convergence of the universe on itself." Teilhard's sense of convergence is similar or identical to coalescence. As the universe or regions of the universe contained within open systems, become more tightly complex and networked or complexified and, thus, convergent, order coalesces. So as the "without" complexifies, not only does consciousness emerge as a property of the "within," but also love emerges. Teilhard writes, "Love alone is capable of uniting living beings in such a way as to complete and fulfil them, for it alone takes them and joins them by what is deepest in themselves" (Teilhard de Chardin 1959, 264–265).

As evidence for love, Teilhard appeals to experience. He points to the depth of feeling in lovers, who "come into the most complete possession of themselves . . . when they say they are lost in each other." He calls attention to the couple or the team (family, village, etc.) in which love personalizes by totalizing. He asks that if love does this on a small scale, might this also be developing on a worldwide scale. As evidence that this may be occurring, he describes an "irresistible instinct" in our hearts that leads us toward unity when our passions are stirred; and we experience a unity with the universe, the "All," for example, by key experiences we may have of nature, beauty, and music. He describes these key experiences of the beautiful and the sublime as "resonance to the All—the keynote of pure poetry and pure religion" and asks, "What does this phenomenon, which is born with thought and grows with it, reveal if not a deep accord between two realities which seek each other; the severed particle which trembles at the approach of 'the rest'?" (Teilhard de Chardin 1959, 264–265).

The cosmic function of Omega, the activity of God attracting the universe toward the future, expressed in the internal (radial) energy of the "within," is to *initiate and maintain* within its (infinite) radius the "unanimity of the world's reflective particles [human individuals]." Here Teilhard applies classical Christian Thomistic theology, which holds that God initiates and sustains existence of *all being* (Davies 1993, 34–36), not just reflective particles. The mode by which Omega accomplishes this, proposes Teilhard, is by love: by the activity of loving and the quality of being attractive, lovable. Thus, Teilhard's position on the nature of God's activity, power, is that it is loving and persuasive as opposed to dictatorial and forceful. A prerequisite for any love is relationship (Teilhard: "coexistence"). Thus, Teilhard holds that while Omega is transcendent, Omega is also immanent—close to the "within" of all things in the universe. Only by being close, intimate, may Omega be "supremely attractive" (Teilhard de Chardin 1959, 269). In this mode of closeness and attractiveness, Omega is located at the nexus of the present and the future and calls, attracts all being on its process toward ultimate completion and fulfillment in the future. Omega is Teilhard's ultimate answer to the escape from entropy that complex interactive networks in open systems achieve and to their tendency toward greater complexity over time (Teilhard 1959, 271). In light of Kauffman and the new sciences of complexity, Omega as an explanation presents itself as a subjective parallel explanation to potentially objective "deep laws of the universe." It addresses questions about our subjective experience as relational persons that impersonal laws, however true, cannot satisfactorily address.

Thus, for Teilhard, the effect of Omega on all living things in the universe is to inspire and initiate the love attraction that "takes them and joins them by what is deepest in themselves." What is deepest in themselves is their "within." In this way, the Omega can only be personal, for as we saw, love personalizes because it is about relationships between personalities. The effect of Omega, for Teilhard, is to spark love-attraction in living

JOSEPH FORTIER

organisms. The effect of a loving relationship brings beings into their own completion and fulfillment, in which human "lovers come into the most complete possession of themselves" and in which, through the experience of great beauty, we sense a "resonance with the All," which brings with it an "awareness of a Great Presence," a revelation of a "deep accord between two realities which seek each other."

It is unclear why Teilhard limits love by definition to living creatures (Teilhard de Chardin 1959, 264). Given his extrapolation of radioactivity further down the periodic table of atoms, from heavier ones in which radioactivity is detectable to those lighter ones in which it can be inferred by extrapolation (Teilhard 1959, 54), it seems he might have described love in the same way. Instead, love is a property of only living things, evidently even bacteria (Teilhard 1959, 264). He does, however, propose that the propensity to unite of the "within" of all things is the primordium of love. But for Teilhard, love is an emergent property only of living systems. It seems that a more comprehensive explanation for the process of complexification within the evolutionary process would be to see the attraction of the Omega as initiating the propensity to unite of the "within" at all levels of organization of matter in the universe, no matter at how simple a level. Teilhard saw consciousness in rudimentary form in the "within" of these simplest of particles. Why is this also not the case for love? Why is love not seen as a constituent element of consciousness even at its most primordial level? It seems that in the spiritual sense, it would need to be the case in order to consistently maintain the position that Omega creates and maintains existence by the attraction of love.

*Emergence of Christogenesis*

Teilhard rarely uses the word *priest* in his writings and never develops it as a concept relevant to his evolutionary Christian theology. Yet *priest* may be seen as a key concept to understanding how he envisions the value of Christian consciousness in relationship with the world in the context of his theological understanding. The concept of priesthood also reveals an important inconsistency in Teilhard's notion of Christian consciousness and the effects of Omega, as we will see. I will approach the task of describing Teilhard's thought about the emergence of Christogenesis, its priestly implications, and Teilhard's apparent inconsistencies by addressing the following questions in the sections below: (i) What is the meaning of *priest*? (ii) What is Christogenesis to Teilhard? (iii) How is immersion in

Christogenesis priestly in a Christian sense? (iv) In what way is Teilhard inconsistent, and how does this frustrate a possible environmental ethic?

a. *What is the meaning of* priest?

In various religious traditions, *priest* is used to refer to an official minister who performs public religious functions (Brown 1993, 2351). In Judaic tradition, the priest or *kohen* was one who cared for the cultic traditions relating to the Arc of the Covenant on behalf of the people (Bright 1981, 200–201). The *kohen* also had other cultic duties, such as sacrificial offerings (Dictionary.com 2009). He may be seen as a *servant-intermediary on behalf of* the relationship between the Chosen People and Yahweh. In Christianity, the priest is one with authority to serve the community by performing certain rites, such as the Eucharist, on behalf of the relationship between the community and God (Brown 1993, 2351).

The New Testament biblical letter of Paul to the Hebrews expounds a theology of priesthood drawn from the Jewish sense of *kohen*, or priest. The letter describes the *kohen* as many through the years, each entering the Holy of Holies in the temple to perform sacrificial rites to Yahweh many times on behalf of the people and serving them compassionately in other ways. In contrast, the letter describes Jesus as the one high priest for all time, having entered the ultimate Holy of Holies (heaven) to make the ultimate sacrifice on behalf of all humans for all time (his own life). His priesthood remains for all time since, having risen from the dead and thus conquering death for us, he lives forever (Hebrews 7:23 to 9:28). Christians are called to participate in Christ's eternal priesthood by how they live their lives, in generous willingness to suffer for the sake of love, to do good works for others, to share material things, all in a sense of participation in Christ's once-for-all-time sacrifice of himself for eternal life (Hebrews 10: 11-14).

b. *What is Christogenesis to Teilhard?*

As we saw, for Teilhard, the noosphere or thinking layer over the earth is a culmination of the process of psychogenesis just as the biosphere, or layer of life over the earth, is a culmination of the process of geogenesis (complexification within the molecular world). Within the noosphere, the blossoming of self-reflective human awareness occurs and continues its process of developing and evolving as the process of hominization or noogenesis. The Christian phenomenon is seen as a blossoming forth of a new kind of consciousness out of the noosphere. Teilhard describes the Christian phenomenon as a phyletic process, also called Christogenesis, in the sense that it is a branch in the process of noogenesis, as an evolutionary process.

The Christian phenomenon is perceived as real and objectively observable by virtue of the influence it has had in all corners of the globe. The Christian faith is practiced in some form in nearly every nation today. Many hospitals and health organizations are either maintained or influenced by Christian organizations or people. Other organizations serve the poor and people in crisis such as the internationally based Church World Service and Jesuit Refugee Service, the US-based Salvation Army, Catholic Relief Services, Catholic Social Services, and Catholic Refugee and Immigration Services. Christian-run schools abound. Unfortunately, most of us are also aware of terrible abuses and scandals visited upon human society by Christian clergy and laypeople, both historically and at present. Teilhard could hardly have been unaware of this since, as we saw, he himself was a victim of such abuse similar to that which Galileo endured. For Teilhard, we're still in process, still victimized by the problem of evil. Perhaps this is why he didn't coin the term *Christosphere* as a blossom of noogenesis. Rather, he sticks with Christogenesis. We continue to deal with what is known in Christian doctrine as original sin, a fallenness that seems timeless, classical in the human condition.

As we saw above, a fruit of the noosphere is the blossoming of the subjective experience of universal love: a sense of the lovability of "the All," of existence for itself. Teilhard describes Christian love as "a specifically new state of consciousness." He highlights its subjective nature, saying that it "is incomprehensible to those who have not experienced it." What is infinite and intangible is experienced as lovable, as are nonhuman beings (Teilhard de Chardin 1959, 295). Christians are called to love their enemies and to do good to those who hate them (Luke 6:27), unfortunately, a call not always heeded. Interestingly, the Hindu holy man Mohandas Gandhi consciously followed this call of Jesus to great effect in the nonviolent resistance movement he led to liberate India from British subjugation. His methods in turn were studied and practiced by the American Christian holy man Martin Luther King Jr. in the movement King led that resulted in the overturning of grossly oppressive racist laws and practices in the United States and the legislation of civil rights laws to protect racial minorities from racist oppression. Christians, however, still initiate and participate in violent wars and are, by their own admission, not perfect followers of Jesus.

Teilhard points to the twenty-century-old practice of Christian mysticism in which people who have committed themselves to the monastic way of life have experienced profound joy in their abandoning themselves to lives of prayer. Trappist monk Thomas Merton's book *New Seeds of Contemplation* (Merton, 1972) offers a taste of this quality of Christian contemplative love in the twentieth century.

Christianity's basic trust in the goodness and lovability of material existence and in the goodness and lovability of the human leads to a certain boldness and inclusiveness of vision that may be unique (Teilhard de Chardin 1959, 296). This vision is inspired by the Christian belief in Jesus as the Jewish Messiah, the anticipated anointed one who was to deliver Israel from the Romans but whose mission came to be seen instead as to deliver all people from the oppression of evil and hate. The observation that we're still on the road inspires fervent, thinking Christians to greater coherence and authenticity in their lived lives and interior faith, greater compassion and appreciation for all their fellow humans, and a sense that they're on the journey for the long haul.

Teilhard was confident that although Christians were "frightened for a moment by evolution," they now see that it offers a "magnificent means of feeling more at one with God and of giving [oneself] more to [God]." He continues that in the view of nature that does not recognize evolution, nature is seen as static and without the coherence and unity that the evolutionary theme of development of all things from common origins and common ancestry sheds light on. But recognition of the evolutionary process in a Christian vision allows the universe to be seen as in a process of coalescing spiritually. If this process of coalescence is the case and if Christ is at its center and also its future goal, attracting all things toward himself, then the process of Christogenesis is "an extension . . . of noogenesis, in which cosmogenesis . . . culminates" (Teilhard de Chardin 1959, 297).

For Christians, the Incarnation, or becoming flesh, of the divinity in the person of Jesus is paramount. It is experienced as a powerfully meaningful expression of God's love and solidarity with all humans and with all creation. In Jesus (the Christ, or Messiah), God became integrated with the earth as a human being, and the dignity of creation and of humanity has become qualitatively more deeply meaningful. Teilhard writes, "Christianity shows itself able to reconcile, in a single living act, the All and the Person. Alone, it can bend our hearts not only to the service of the tremendous movement of the world which bears us along, but beyond, to embrace that movement in love" (Teilhard de Chardin 1959, 298).

*c. How is immersion in Christogenesis priestly in a Christian sense?*

As we saw, a theology of Christian priesthood drawn from Paul's letter to the Hebrews is about solidarity with Jesus, who is believed by Christians to be the divine Son of God (John 1:1–18, 20:24–30) and in whom God became in solidarity with all humanity and the universe by the Incarnation. Since Jesus is seen as the once-for-all-time high priest who offered himself as the sacrifice for renewal of creature-God relationship and who has come to live eternally, the Christian community experiences itself as participating in Christ's once-for-all-time priesthood. All Christian-ordained priesthood is seen as performing priestly sacramental and other cultic duties in the priesthood of Christ, on behalf of the priesthood of all Christians.

Given that Christogenesis is a process of becoming, of moving toward a future fulfillment, and given the evidence for the imperfection of Christians in living their faith, it's obvious that Christian participation in the priesthood of Christ is a work in progress. The major subjective insight essential for Christian priestliness is that the Incarnation (becoming flesh) of the divine in Jesus and Jesus's life, death, and resurrection from the dead *was God's expression* of his love for the world, including humans. Insofar as this Incarnation is understood as the spirit of the divine becoming more intimately joined with the "within" of the All, using Teilhard's language, Jesus can be understood as the one high priest in the Judeo-Christian sense. Since this awareness brings to Christians, as participants of the noosphere community, a subjective interior movement of great love for God and for all that God loves (the All), they become interiorly disposed as agents of Christ's priesthood. As they are moved by this great love to perform deeds of love for one another and for all people, as well as for all material existence, including promotion of just respect and acknowledgment of the dignity inherent in all people and nonhuman nature by virtue of their "within" relationship with God, they live out their priesthood. Thus, to the extent that Christians express aware love on behalf of all creation for God's solidarity with all creation, they are priestly, experiencing inwardly and acting on behalf of the love relationship between God and all creation. This is expressed biblically in the book of Paul to the Romans (Romans 8:18–23), which describes the "longing" and "groaning" of all creation for the fulfillment of the human relationship with God (the process of Christogenesis).

## d. *In what way is Teilhard inconsistent?*

After developing a monistic description of the natural history of the earth as a process of dynamic relationship in which life, consciousness, and reflective awareness emerge, Teilhard seems to renege on this monism when he addresses humanity's final end. The essential aspect of Teilhard's monism is the integrity of matter and spirit as one. The simplest organization of matter and energy of which the universe is built is the atom. For Teilhard, each atom is a dynamic relational unity composed of an exterior and an interior aspect, capable of interrelating with other atoms. As aggregates of atoms complexify in their interrelationships, new layers of complexity-consciousness emerge over the earth (biosphere, noosphere). Complexity of the organization of matter and increase of consciousness emerge in tandem as matter and energy complexify in their interrelationships over time: the law of complexity-consciousness (Teilhard de Chardin 1959, 61–66; 1970, 155–157).

It would seem logical that any final culmination of this process of complexity-consciousness, which includes the noosphere and Christogenesis, would not compromise the integrity of the monism of matter and spirit. On the contrary, especially with the emergence of love, universal love to wit, it would seem that the universe comes as a package. A problem with self-reflective awareness that Teilhard observes is the problem of death. The "death-barrier" discourages the hopes and thus the interior dynamism, zest for life in the self-reflectively aware person (Teilhard de Chardin 1970, 397). The solution Teilhard provides is the traditional Christian solution, eternal life for the human particles of the noosphere, in which the conscious converges on itself and the noosphere splits from geogenesis (Teilhard 1959, 287, 289; 1970 399, 402–403).

Thus, Teilhard's solution to the problem of death and the universal human sense of eternal life (Teilhard de Chardin 1970, 402–403) is the splitting off of the human "within" from the rest of creation. Dualism, or segregation of matter and spirit, creeps back into the picture. A key question is begged: why would a loving God who, according to Christian conviction, became one with the universe in matter and spirit in the person of Jesus and ascended into eternal life in body and spirit, not in some way share this same new life with not only human souls but also with all his beloved creation? In fact, Teilhard seems to ignore a tenet of traditional Christian belief by not addressing the bodily resurrection of the dead in the future. For Teilhard, radial energy, a relational property of the "within" of things, is the glue of the universe and finds its ultimate center at the Omega point of the universe. The Omega inspires the relational aspect of all things in the universe, out of which, by the law of complexity-consciousness, love has emerged in living creatures. The activity of God is love, as Teilhard recognizes. Love becomes fully reflectively conscious in human persons. Love is a dynamic, mutual relationship. Thus, a difficulty with Teilhard's cosmogenesis seems to be that he doesn't develop the relationship of the "within" and love or at least explicitly consider "within" as a primordium of love. To do so would seem a more parsimonious explanation of the integrity of the universe and its relationship with God than an *independent* emergence of love by evolution from the primordial "within" after the threshold of life had been crossed, as Teilhard does. To recognize love in primordial form in the "within" of all things would also strongly suggest, for a Christian evolutionary theology, the share of the entire universe in Christ's resurrection. This would seem necessary for the integrity of Teilhard's monism.

Finally, in order for Christogenesis or Christian universal love to find its completion and fulfillment, there is the priestly vocation of the human in the universe to consider. If Christian consciousness allows us to "reconcile, in a single living act, the All and the Person," to "bend our hearts not only to the service of the . . . movement of the world" and to "embrace that movement in love" then is it not allowing us to understand our solidarity with the world, God's solidarity with it, and its lovability to God? Might an alternative to Teilhard's separation of the noosphere from geogenesis be the eternal integrity of matter, in its monism of exterior face and spiritual, relational "within"? Teilhard's ultimate dualistic solution seems to compromise Christian participation in Christ's priesthood with respect to the rest of the universe.

In the following chapters, we will explore further how Christian thought is discovering how the science of evolution deepens and informs the content and meaning of Christian faith and how the conception of Omega calls on advocates of the science of evolution to see that there may be more to evolution than meets the empirical eye.

# Summary

☐ Teilhard de Chardin was a Jesuit priest and paleontologist during the first half of the twentieth century. His teaching faculties were taken from him by authorities in the Catholic Church and Jesuit order because of his advocacy for evolution and his doubts that Adam and Eve were actual historical people. He nonetheless remained a Jesuit priest, paleontologist, and writer throughout his life. His writings were published after his death in 1955 and are considered to have been influential in the Second Vatican Council of the Catholic Church.

☐ Teilhard developed a monistic position with respect to the universe; that is, that the basic substance of the universe is one. In doing so, he rejected dualism, or the dividing of reality into matter and consciousness. Instead, he saw the basic unit of the universe, the atom, as having both an outward aspect, objectively knowable and measurable, and an inner aspect, or "within," which is subjective and not knowable by empirical methods.

☐ Over time, as the universe and in particular the earth has evolved, a process of complexification of relationships among atomic and molecular particles and living particles composed of these basic particles has coincided with the rise of consciousness from the primordial protoconsciousness of the atomic and molecular "within." This tendency for consciousness to increase with the increase in the outwardly observable complexity of relationships among constituent particles of a thing is called the law of complexity-consciousness.

☐ Some properties of the universe only become evident at large scales of measurement of other coincident properties. For example, radioactivity only becomes evident in large atoms with high molecular weights, which, when first discovered, was a clue that radioactivity is given off, though at trace amounts, by all atoms. In the same way, consciousness only becomes evident in aggregations of matter with high complexity (diversity of kinds of atoms and molecules and of relationships between those constituent atoms and molecules), such as living creatures. Thus, we may infer very dim consciousness in simple organisms and aggregations of matter with low complexity.

JOSEPH FORTIER

- [ ] The entire universe is an interwoven net of interrelational complexity and is itself a monad.
- [ ] There are two kinds of energy in the universe: tangential energy and radial energy. Tangential energy is observable and measurable. Radial energy is not. Radial energy is the energy of the "within" and draws an entity in the universe toward greater complexity or centricity (the capacity of an entity for more conscious relationship with other entities).
- [ ] Teilhard's ideas are comparable to Kauffman's science of complexity. Both deal with laws, and both describe the process of growing complexity in the universe as an important component of evolution. In both sets of ideas, crossing thresholds of complexity or critical points lead to the emergence of new order with new properties that could not be predicted before the actual emergence.
- [ ] Teilhard anticipated four qualities in Kauffman's empirical description of emergent order in living systems over the evolutionary process:

  1. The indefiniteness of what will emerge when an ordered system is poised at the edge of chaos.
  2. The fixity, or robust resistance to substantial change in basic pattern of the newly emergent order.
  3. The flexibility of the newly emergent order allows it to evolve in a complex branching system of diverse varieties while each branch of the evolving system maintains the basic integrity of the pattern in the order. Thus, an embryonic stem cell may develop into one of dozens of kinds of particular cell types.
  4. When thresholds of complexity or critical points of the complexity among diverse kinds and numbers of particles and their interrelationships are crossed, new order condenses or collapses into place.

- [ ] Teilhard anticipated Eldredge and Gould's punctuated equilibrium theory when he described the successive evolution of emergent, newly complex order as occurring in jumps.
- [ ] Teilhard's scenario for the evolution of life on earth incorporated the idea of complexity-consciousness in the process of geogenesis. Geogenesis, the process of complexification of nonliving, mineral matter, led to the emergence of life: the biosphere, or layer of life

over the earth. The process of psychogenesis is that as living systems complexify, at each step in the jumps of phyletic complexification, the "within" of the system becomes a bit more conscious, capable of some progressive degree of awareness, of some progressive degree of ability to relate in a more complex way. The process of psychogenesis leads to the major threshold over which the noosphere emerges: the human capacity for self-reflective awareness, complex language, and awareness of the transcendent. From within the noosphere, the process of noogenesis emerges; and various cultures, languages, sciences and technologies, and religions emerge. One branch of noogenesis is the Christian phenomenon: Christogenesis.

□ An aspect of noogenesis is the capacity to experience universal love: the lovability of something beyond, of the All. The All seems to include the universe and the Omega point of the universe: that center of attraction of God's love.

□ Christogenesis is the development, after the Christ event (the Incarnation of God in the person of Jesus and his death and resurrection), of this universal love. Christogenesis develops as focus on the attractive Omega point draws the human particles of the noosphere toward itself, into the future.

□ Love is a phenomenon that can only be experienced subjectively and is a fruit of the process of complexity-consciousness through psychogenesis and especially noogenesis after having crossed the critical threshold into the noosphere. In its reflectively conscious, free form, it is a manifestation of the human "within," or soul.

□ By virtue of the reflective consciousness and for the capacities for universal love and for relationship with the transcendent or Omega, the human particles of the noosphere are priestly in that they mediate the relationship of God's love with all the rest. Teilhard unfortunately moves away from this priestly role of the human in its relationship with God and with the rest of the universe instead of developing it when he addresses death and, thus, compromises his idea of the universe as a monad.

□ Although Teilhard equates the "within" as a primordium of consciousness, mind, he does not equate it as a primordium of love, soulfulness. Yet he describes love and soul as emerging in the psychosphere after the threshold into the biosphere has been crossed. This lack of continuity of love and soul with "within" begs explanation.

# Glossary

**biosphere**. The relatively sudden emergence of living organisms from prebiotic soup, and the growth of life into a layer over the earth.

**centricity**. Degree of capacity for conscious relationship.

**Christogenesis**. A branch of the phyletic process of hominization in which the Christian phenomenon emerges and grows in its realization of the cosmos' and humanity's essential dignity and lovability because of the presence of Christ in the world.

**complexification**. The aspect of the evolutionary process by which newer lineages emerge that are qualitatively more complex than older lineages of organisms. The major complexification events occur in jumps.

**complexity-consciousness**. The aspect of the evolutionary process by which, as complexification occurs, the "within" also becomes progressively more conscious.

**critical point**. The tipping point at a threshold of change in physical state of a substance.

**hominization**. The process of the growth and development of the noosphere.

**Incarnation**. The event in which God became human and part of the ecology of the earth in the person of Jesus.

**monad**. An adjective for describing the universe or its entities as profoundly one with material and spiritual aspects.

**noosphere**. Result of a giant evolutionary jump within psychogenesis resulting in a reflective conscious layer over the earth at the top of the living layer (biosphere): uniquely human reflective consciousness capable of being articulated with highly developed language within a social context.

**Omega point**. A center of the noosphere and of the universe, existing in the future in its transcendence and at the nexus of the present passing into the future in its immanence, at which all centers of everything in the universe converge and become completed and fulfilled. A point at which the loving god is in the universe, and all of it is in God. The effectiveness of the Omega point lies in its attractive, loving, noninterfering activity on the subjective "within."

**orthogenesis**. A theory, largely abandoned, that new kinds of organisms (descendants) in some instances do not branch from older kinds of organisms (ancestors) but rather emerge at the time of death of the ancestral kind so that species succession occurs in a linear fashion, without branching of new lineages from ancestral ones.

**panpsychism**. A philosophical position that holds that all entities in the universe have some degree of consciousness, even in rudimentary form.

**psychogenesis**. Biological process of the rise of the ability to learn by some organisms over evolutionary time.

**radial energy**. Energy which is not empirically quantifiable and which draws the entity toward increasing complexity and centricity.

**tangential energy.** Empirically accessible energy that links entities such as atoms and molecules with others of the same order of complexity and centricity.

**threshold of complexification**. A stage in the process of biological evolution at which the conglomeration of particles in a system growing in complexity crosses some critical threshold of complexity, interrelationships intensify, and the conglomeration becomes a qualitatively more highly ordered unit. Examples are the emergence of life from nonlife and the emergence of the ability to learn in complex living organisms.

**within**. The subjective, interior aspect of every entity in the universe.

# References

Bright J. 1981. *A History of Israel.* Philadelphia PA: Westminster Press.

Brown L. (ed.) 1993. *The New Shorter Oxford English Dictionary.* Oxford: Clarendon Press

Campbell N. A., J. B. Reece, and E. J. Simon. 2007. *Essential Biology, Third Edition.* San Francisco, California USA: Pearson/Benjamin Cummings.

Davies B. 1993. *The Thought of Thomas Aquinas.* Oxford, Great Britain: Oxford University Press.

Dictionary.com. 2009. Kohen. Last modified 2011. http://dictionary.reference.com/browse/kohen

Eldredge N. and S. J. Gould. 1972. "Punctuated equilibria: an alternative to phyletic gradualism." In *Models in Paleobiology,* edited by T. M. Schopf, San Francisco, CA: Freeman, Cooper & Co.

Futuyma D. J. 2005. *Evolution.* Sunderland, Massachusetts, USA: Sinauer Associates, Inc.

Gould J. 1989. *Wonderful Life: The Burgess Shale and the Nature of History.* New York: W.W. Norton & Co.

Gould J. 2007. *Punctuated Equilibrium.* Boston, MA: Belknap Press.

Gould S. J. and N. Eldredge. 1977. "Punctuated equilibria: the tempo and mode of evolution reconsidered." *Paleobiology* 3: 115-51

John. 1977. "The Gospel According to John." In *The New Oxford Annotated Bible.* May B. M and H. G. Metzger (eds.). New York: Oxford University Press

Kauffman S. 1995. *At Home in the Universe; The Search for the Laws of Self-Organization and Complexity.* New York: Oxford University Press.

Kauffman S. 2008. *Reinventing the Sacred: A New View of Science, Reason, and Religion.* Philadelphia, Pennsylvania USA: Basic Books.

Lukas M. and E. Lukas. 1977. *Teilhard.* Garden City, New York: Doubleday and Company, Inc.

Luke. 1977. "The Gospel According to Luke." In *The Oxford Annotated Bible.* May B. M and H. G. Metzger (eds.). New York: Oxford University Press

McGrath A. 2006. *Christian Theology: An Introduction.* Oxford, Great Britain: Blackwell Publishers.

Paul. 1977. "Letter of Paul to the Hebrews." In *The New Oxford Annotated Bible.* May B. M and H. G. Metzger (eds.). New York: Oxford University Press.

Roach J. 2008. "Oldest Primate Fossil in North America Discovered." *National Geographic News*, March 3. http://news.nationalgeographic.com/news/2008/03/080303-american-primate.html.

Scigliano E. 2003. "Through the eye of an octopus." *Discover*, October 1. http://discovermagazine.com/2003/oct/feateye/?searchterm=through%20the%20eye%20of%20an%20octopus.

St. John E. 2008. *Action, reflection, and social justice: Integrating moral reasoning into professional development.* Cresskill, New Jersey USA: Hampton Press.

Teilhard de Chardin P. 1959. *The Phenomenon of Man.* New York: Harper & Row, Inc.

Teilhard de Chardin P. 1970. *Activation of Energy.* London: William Collkins Sons & Co. Ltd.

Teilhard de Chardin P. 1971. *Christianity and Evolution.* New York: William Collins & Sons, Inc.

Thesaurus.com. 2009. "Complex." http://thesaurus.reference.com/browse/complex?qsrc=2889

# CHAPTER 8

# What is Evolutionary Theology, and How Does it See Darwinian Evolution as Gift?

I N THIS INTRODUCTORY chapter to evolutionary theology, we'll look at what evolutionary theology is and how the science of evolution informs Christian theology. Notice that this isn't about proving theological points, like the existence of God or the necessity for God's existence, by using scientific evidence. It is about entering into a disciplined reflection on how Darwinian evolution informs and deepens human understanding of who God is and the timeless significance of Jesus to Christian faith.

## What is evolutionary theology?

Evolutionary theology is not natural theology, which looks to science for evidence of God's existence and nature (Barbour 2000, 3, 28). Rather, it is a theology of nature, which assumes a certain religious tradition and investigates how some of its beliefs should be reformulated in the light of scientific discoveries in order to come to a more consonant vision of truth (Barbour 2000, 3). Theologies of nature do not deny the universal human perception of "Great Mystery" to use the English translation of the Lakota Sioux term *Wakan Tanka* since they acknowledge the legitimacy of subjective human experience as a means for approximating toward the truth of things.

Intelligent design theory, which is an antievolution religious position dressed as science, attempts to interpret certain scientific facts as evidence for irreducible complexity in the sense that complex forms could not have arisen from less complex forms by the process of evolution. In 2005, in the case *Kitzmiller et al. v. Dover Area School District*, Judge John E. Jones found that intelligent design theory is not science but a creationist religious position that falsely claims to be science (Jones 2005). Intelligent design theory is an example of applying natural theology in an intellectually dishonest way. It attempts to find evidence for an intelligent designer by using science in a way that manipulates and misconstrues scientific findings in order to present "evidence" that would negate evolution. Not one publication supporting intelligent design has ever been legitimately published in a scientific journal, although in 2004, lax peer review allowed one through the cracks. The article was shortly thereafter retracted by the journal on grounds of lack of scientific rigor (National Center for Science Education. 2004).

In contrast to intelligent design theory, evolutionary theology explores how new awareness of cosmic and biological evolution as disclosed by the scientific community can inform and enrich traditional Christian teachings about God and God's activity. Rather than regarding evolution as a dangerous challenge that calls for a response, evolutionary theology finds in evolution an illuminating context for seeking understanding about God and matters of faith (Haught 2000, 36).

## What does evolutionary theology say about creation?

Traditional Christianity has recognized three aspects of God's creative activity: original creation (*creatio originalis*), ongoing, continuous creation (*creatio continua*), and fulfillment of creation in newness (*creatio nova*). Before Darwin, the aspect of original creation generally dominated thinking about God's creative activity. At present, with the development of big bang theory, discussions between scientists and theologians generally assume the original creation aspect. However, even by the end of the nineteenth century, some theologians recognized that the process model that evolution brings to the table liberates thinking about creation from the issue of cosmic origins only. Today, thinkers such as John Haught find that the idea of evolution forbids the constriction of so powerful a notion as divine creativity to simply the one aspect. Instead, the discoveries of evolution allow theology to broaden its vision and to see God's creative activity as an ongoing, forever new reality. Creation is still happening now, just as "in the beginning" (Haught 2000, 37).

Since the universe is still in process, not finished, then we cannot expect it to have reached perfection. If we live in an imperfect world still in process, then the existence of suffering and evil is conceivable, even to a theistic position that understands God as good and loving. However, that being said is not to say that such phenomena are tolerable, nor that they are God's will, but rather that they are the effect of the incompletion of the creative process (Haught 2000, 38). Since Christian faith is a way of hope, embedded within it is the notion of progress. Since it sees moral value in all things (Genesis 1:31), it hopes for the completion and fulfillment of creation (*creatio nova*), which would mean an end of suffering and evil.

## What does evolutionary theology say about the future?

Theologian John Haught takes a different tack from Teilhard with respect to Christian hope and the future. He isn't as focused as was Teilhard on directional evolution, or orthogenesis, with humanity (the noosphere) at the top of a hierarchy of emergent order. Like Teilhard, he sees a future-oriented, evolutionary context for human hope that would place our hope "outward and forward." Unlike Teilhard, his hope is located in solidarity with the entire evolving cosmos in the context of its ongoing creation. In doing this, we retrieve an often overlooked biblical theme that envisions the universe as included in the final fulfillment. For example, Paul describes all creation groaning for ultimate fulfillment (Romans 8:22–23). Certainly, this notion is consonant with the post-Darwinian awareness that cosmic and human destinies are joined.

## What does evolutionary theology say about grace?

In Christian thought, grace is the activity of God's loving kindness in the world. How do we square this with the fact of suffering, the merciless competition of natural selection, and seemingly impersonal, uncaring randomness? To answer this question we need to take a glimpse at the nature of love. Authentic love doesn't control, force, or dictate. The subject of authentic love respects the dignity of the beloved. It cannot simply control. Rather, it feels compelled to suffer with the developmental process of the beloved. It doesn't attempt to make the beloved an appendage of itself. Rather, it longs for the beloved to become completed, fulfilled in its unique self, in its own way. Although love is nurturing, compassionate, and seeks to understand, it also longs for the independence and freedom of the beloved. In order for there to be authentic relationship with a beloved then, the intimacy between the lover and beloved must be that of dialogue with *the other*. There must be differentiation. Thus, God's grace must mean that God lets the world be and become itself. If this is so, then it may seem that God's love takes the form of reserved self-withdrawal so that the world may emerge more autonomously and, thus, allow for an authentic relationship with God. Thus, the natural history of the world is the story of its evolutionary odyssey through eons of time, of relationship building upon relationship, of struggle, destruction, building, competition, cooperation, suffering, and joyfulness—within the context of expansive freedom in the presence of self-giving grace (Haught 2000, 39–40). The issue of God's love and its effectiveness will be revisited later.

## What does evolutionary theology say about God's power?

According to the process philosophy and theology pioneered by Alfred North Whitehead, evolution informs us that God's power in relation to the world is expressed in the action of persuasive love rather than coercive force. Consistent with the freedom of God's grace, God's love invites and does not compel. The question may be asked whether a policy of persuasiveness rather than coerciveness in relationships is a sign of weakness. If God is about love and if God's objective is to inspire love in the creation, then coerciveness would be much less apt to nurture a world in such a way that creatures capable of freely chosen love emerge than would persuasive love. The capacity for truly deliberative love might never evolve in the puppetlike universe of a coercive, controlling deity. Unfortunately, immature religiosity exists, which often wishes for such a coercive, designing deity, and this seems to be what scientific skeptics have in mind when they attempt to assert that Darwinism and modern science have placed the lid on the coffin of belief in God (Haught 2000, 40–41; Kauffman 2008, 283). The power of God's love is discussed in more detail below, under "How is Darwinian evolution a gift to Christian faith?"

## What does evolutionary theology say about the meaning and value of existence?

Again, Haught's evolutionary theology draws upon Whitehead's process thought and finds a feeling, relational God who responds to the world of evolution. For Whitehead, God, in order to authentically relate with the relational creation, on some level enters into the process of ongoing existence. While remaining transcendent, God also remains immanent, in solidarity with the process of existence.

On the surface, from a purely objective view through a particular lens, this world seems to be a world full of beauty and also full of constant meaningless, pointless suffering and dying. Yet numerous biblical passages portray God's responsiveness to the world, God's feeling for the world, and God's desire to care for the world, as well as God's creativity. These themes are also found in other religious traditions. From the perspective of a cosmos undergoing the process of evolutionary becoming, such a god, Haught reflects, would seem to be deeply affected *on a subjective level* by all happenings throughout the history of the evolutionary process. Thus, all events in the course of this history with their suffering, tragedy, as well as the emergence of new life and deep beauty, enter into God's own subjective, feeling experience of the world. Thus, in one sense, all events and accomplishments in the natural history of the world are ephemeral, perishable. Yet in another sense, they always remain within the eternal compassion of God's memory and, thus, never perish in an absolute sense. Thus, the meaning of their existence is never extinguished. The vagueness and darkness of our understanding of things of such ultimate meaning is consonant with the unfinished, still-in-process state of the universe. The nature of God's infinite, active love, also consistent with this view, gives the aware person a reason to hope (Haught 2000, 43).

## How is Darwinian evolution a gift to Christian faith?

Teilhard's pioneering work in developing a position that integrates cosmic and biological evolution with Christian faith focuses on the emergence of complex systems such as the rise of complexity in organisms that biology investigates and the concurrent rise of consciousness with increasing biological complexity. His work suggests an interior subjectivity at the heart of matter not accessible to empirical investigation but arising as consciousness over evolutionary time in a way proportional to the exterior, objectively observable rise in complexity. However, his work does not specifically address the consequences of Darwinian evolution as such: a world of randomness, contingency, competition, suffering, and seeming purposelessness. Thus, Teilhard does not address the Darwinian invitation to face squarely the problem of theodicy: the dilemma of apparently meaningless suffering and cruelty in what is purported by Christian faith to be the good creation of a loving, good god.

For John Haught, theological investigations that only look for design and order, even the order that condenses from the complex networks of Kauffman and Teilhard, fail to address the religious sense of divine pathos. For Haught, any description of God's relationship with the cosmos as only that of a sort of unmoved mover located in the future, attracting all by persuasive love, does not fully address the passionate commitment of God to relationship with this cosmos. The at times irrational passion of God expressed in scripture and the anguished intimacy of God's ancient and ageless affair with the cosmos is not sufficiently addressed by such investigations (Haught 2000, 46). At the same time, Haught places the Darwinian view of life in a larger, more existential context that simply the materialist context assumed by such conflict-oriented thinkers as Phillip Johnson of the intelligent design school and Richard Dawkins of the scientific atheist school. With Teilhard, he addresses the subjectivity of living things. The chance and contingency of mutations, the workings of natural selection, and the vast timescales in which life's evolution unfolds are not adequately described by mechanistic concepts alone. For Haught, the data of evolutionary science can be more completely and fully understood when also situated within the context of the biblical humility of God (Haught 2000, 47).

A challenge for those who are committed to the divinity of Jesus and the timeless significance and power of his suffering and death is to see the true effectiveness of God's power in Jesus's vulnerability. Through Christianity's teaching of the three persons sharing in the one Godhead (the Father, Son, and Holy Spirit), it clearly recognizes the crucifixion of Jesus the Son as an internal aspect of God's experience rather than as an event external to God. Unfortunately, the scriptural and early Christian doctrinal sense of God as "suffering servant" faded with the legitimization of Christianity by the Roman Empire under Emperor Constantine. The image of Caesar replaced the image of the humble shepherd of Nazareth as the predominant way of understanding God. Unfortunately, it is this image and understanding of God that hovers over the heads of the intelligent designer creationists and evolutionary materialists as they continue their tedious and ultimately pointless debates (Haught 2000, 48).

Unfortunately, some writers have described the god of Jesus as "powerless" and "weak." However, Edward Schillebeeckx makes the distinction between lack of divine power on one hand and divine participation in the world's suffering on the other. Thus, the image of God's humility and suffering love implies a sort of defenseless vulnerability, which can powerfully

JOSEPH FORTIER

and effectively disarm evil, rather than weakness and powerlessness. Schillebeeckx writes that human notions of power, with their capacities for destructiveness, have nothing in common with divine omnipotence since that omnipotence expresses itself in this world as defenseless vulnerability. It reveals itself as the power of love that challenges human power structures and offers life and freedom (Haught 2000, 48–49).

Mohandas Gandhi, who led the liberation of India from British colonial rule through a campaign of nonviolent civil disobedience with the underlying principle of suffering love for the opponent, said, "The name of Jesus at once comes to the lips. It is an instance of brilliant failure. And he has been acclaimed in the west as the prince of passive resisters. I showed years ago in South Africa that the adjective 'passive' was a misnomer, at least as applied to Jesus. He was the most active resister known perhaps to history. His was non-violence par excellence." (Kripalani 1970).

Martin Luther King, a Christian religious authority in the Baptist denomination who led the American civil rights movement in the early 1960s, studied Gandhi's spirituality of Jesus and nonviolent resistance methods. On various occasions, he said, "At the center of non-violence stands the principle of love," and "Have we not come to such an impasse in the modern world that we must love our enemies—or else? The chain reaction of evil—hate begetting hate, wars producing more wars—must be broken, or else we shall be plunged into the dark abyss of annihilation." During the nonviolent civil rights demonstrations of the 1960s, King and his followers were beaten, jailed, fire-hosed, and threatened with death by the enemies they worked to love—and maintained their disciplined nonviolent activism on behalf of people of color in the United States. In 1964, the American Civil Rights Act was passed, which outlawed racial segregation in schools, public places, and employment. In 1968, King was assassinated, as was Gandhi in 1948. Although these leaders suffered and were killed, they, like Jesus, were not weak men. Although God's power is revealed in defenseless vulnerability, it is not weak.

If the power of God, as it appears to us, can be characterized as humble, willing to suffer to build relationship based on love, and bearing the fruits of life with all its dynamic spontaneity and freedom, then we must ask what this says about original creation (*creatio originalis*). Jurgen Moltmann, a prominent twentieth-century theologian best known for his theology of faith and hope in the context of suffering, wrote that God's creative activity is preceded by his "humble self-restriction." In this context, God's creative activity, both in the sense of original creation of existence and in that of the ongoing creation of the cosmos, can be characterized as an activity of the self-emptying love of a helping servant (Haught 2000, 49).

Thus, a theology that addresses the evolutionary process of being and becoming, with its subjective inner face of senseless suffering and tragic death along with its joy and pleasure, must be responsive to these subjective effects of the objectively observed spontaneity and random freedom that scientific evidence finds in this process. Such a theology, informed by reflecting on these aspects of the evolutionary process, finds that the nature of God is humbly, lovingly self-restricting, and allows the existence that he ushered forth freedom to become itself. This theology finds that God maintains his consistency in this humble, self-restrictive mode of serving so that the creation is nurtured by loving presence, allowed its freedom to persist in its own process of being and becoming, influenced but not controlled by its humble servant-creator. The strength of God, then, is not revealed in a relationship defined by forcefulness, but rather in one characterized by persistent, long-suffering love.

According to the ancient Chinese wisdom tradition of Taoism, the ultimate reality, or Tao, is conceived as being energetically passive but informationally active. In what at first seems paradoxical, the inactivity of Tao initiates the structure and function of the world. The activity of Tao, or *wu wei*, is perhaps best interpreted as "noninterfering effectiveness." The *Tao Te Ching*, a literary work attributed to Lao-tzu and written in the sixth century BCE, describes the Tao as like water or a valley. These things accomplish much while remaining unobtrusive and passive. This Tao that informs nature is so withdrawn from empirical accessibility that it cannot be named. It is characterized by noninterference yet is powerful in its self-withdrawal (Haught 2000, 78).

Wu Cheng, the Taoist philosopher who lived from 1249 to 1333, described the Tao as like the space in the hub of a wagon wheel, which, by its empty presence, makes movement of the wagon over the ground possible. Likewise, the hollow space in a bowl makes it possible for the bowl to store things; and the emptiness of a room, as well as the space of doors and windows that allow light in, make a place to live possible (Haught 2000, 79).

JOSEPH FORTIER

# Synopsis

To synopsize, Darwinian evolution is a gift to Christian faith in that it challenges Christians to rethink their notions of God's power and activity in light of the randomness and seeming purposelessness of life in the context of the process of evolution perceived from an objective, empirical, mechanistic viewpoint, and in light of the pain and suffering in the world, to name one aspect from a subjective viewpoint. The gift given to Christianity is that of recognizing the subjective nature of God's power. God's power is not about the forces on which physics sheds its light but rather about the subjectively experienced humble, persistent, suffering love of God in solidarity with the process of existence that is expressed in the life and death of Jesus. The strength of this self-giving, serving love is revealed in how great, strong leaders such as Gandhi and King, inspired by Jesus's life and teachings, followed his example by their own disciplined, nonviolent, suffering love for their fellow humans. Since God's power is about nurturing love rather than forceful control, the Christian vision of God's relationship, also informed by Chinese wisdom, understands God's power as being energetically passive while informationally active, in a way that is powerfully effective. God's power may be seen as informing the cosmos and its evolutionary processes of becoming, as Teilhard saw it, by infusing it with his love. However, we are informed by the manifestation of God's love in the life of Jesus that this love is not only transcendent above and beyond, but is also within. God is very close, suffering and feeling along with the travail of the evolutionary process, informing all things by his self-giving, nonobtrusive presence. Teilhard's subjective observation of a "within" of things that would connect with other "withins" by a sort of subjective energy not accessible to empirical observation is tremendously helpful in understanding the subjective nature of God's creative activity in a free, self-determining, relational universe governed by God's nonobtrusive, loving power. To paraphrase John Haught, the logic of both Taoism and a theism of a God of self-giving, suffering love deflates the empirical demand that everything real must make itself available to physical, sensate observation (Haught 2000, 80).

# Summary

☐ Evolutionary theology is a theology of nature in that it assumes a religious tradition and investigates how scientific investigation of evolution can inform understanding of that tradition so that its articulation can be refined to better approximate the truth.

☐ Intelligent design theory is an antievolution, creationist, religious position that the 2005 *Kitzmiller et al. v. Dover Area School District* found to not be science but rather a creationist, religious position falsely claiming to be science.

☐ Evolutionary theology recognizes the three traditional Christian aspects of God's creative activity: original creation of existence, continuous maintenance of the creation, and the future fulfillment of creation. Before Darwin, the original creation aspect dominated Christian thought about God's creative activity. However, since then, thinkers such as Pierre Teilhard de Chardin and John Haught see God's creative activity as an ongoing, dynamic, forever-new reality.

☐ Evolutionary theology brings a future-oriented view to the evolutionary process in which God's creative activity is active. This future-oriented view of God's creative activity is a cause for hope in the final fulfillment of the entire evolving cosmos, including humans.

☐ With respect to the notion of grace, or God's loving kindness, evolutionary theology understands God's love as respecting and nurturing the growth and development of the cosmos in its evolutionary process without controlling it. In this way, the differentiation that develops between God and the cosmos allows for authentic relationship to develop with God in the context of God's grace.

☐ With respect to the notion of God's power, evolutionary theology understands the expression of this power as persuasive love rather than coercive force. Coercive force would be much less apt to bring about a world in which the capacity for love emerges, such as it has, than the power of persuasive love.

☐ With respect to the meaning and value of existence, God is understood by evolutionary theology as a feeling, relational, personal God who responds to the world in its evolutionary process. While maintaining God's transcendence, God is also immanent, in close solidarity with the process of existence. In his subjectivity, God is affected by the events that occur in the history of the world; and all these events, with their suffering, tragedy, beauty, and joy, are retained in the eternal memory of God.

☐ The Darwinian view of evolution is a gift to Christian faith precisely in that it asserts the randomness, contingency, competition, suffering, and seeming purposelessness in the world. The credibility of notions of God's power that have to do with control is foiled by these realities and challenge Christian thinkers to see the true effectiveness of God's power in Jesus's suffering and death. The image of God's humility and suffering love in his relationship with the world implies a quality of defenseless vulnerability, which, rather than showing weakness, can powerfully and effectively disarm evil by appealing to the capacity of the universe to respond to divine love. Thus, the gift of Darwinian evolution to Christian faith is that it challenges Christian thought to question its ideas of perfection and power inherited from Greek philosophy and instead return to its core faith in God's suffering love as revealed by Jesus as the power that moves the universe.

☐ A parallel, complimentary way that the power of God can be seen, described by the Chinese Taoist tradition, is the *wu wei*, interpreted as noninterfering effectiveness. The *wu wei* can be described as the activity of being or providing a space that allows for the activities of events and entities.

# References

Barbour I. 1999. *When Science Meets Religion*. San Francisco, CA: HarperSanFrancisco.

Haught J. 2000. *God After Darwin: A Theology of Evolution*. Boulder, Colorado USA: Westview Press.

Jones J. E. 2005. Tammy Kitzmiller et al., Plaintiffs v. Dover Area School district et al. Defendants (4:04-cv-02688-JEJ). District Court for the Middle District of Pennsylvania. http://www.google.com/search?q=2005.+Tammy+Kitzmiller+et+al.%2C+Plaintiffs+v.+Dover+Area+School+district+et+al.+Defendants+%284%3A04-cv-02688-JEJ%29.+District+Court+for+the+Middle+District+of+Pennsylvania.+&ie=utf-8&oe=utf-8&aq=t&rls=org.mozilla:en-US:official&client=firefox-a

Kauffman S. 2008. *Reinventing the Sacred: A New View of Science, Reason, and Religion*. Philadelphia, Pennsylvania USA: Basic Books.

Paul. 1977. "Letter of Paul to the Romans." In *The New Oxford Annotated Bible with the Apocrypha*, ed. May H. G. and B. M. Metzger. New York: Oxford University Press.

National Center for Science Education. 2004. "BSW Strengthens Statement Repudiating Meyer Paper." Last modified 2011. *http://ncse.com/news/2004/10/bsw-strengthens-statement-repudiating-meyer-paper-00528*.

JOSEPH FORTIER

# CHAPTER 9

# Toward an Ultimate (theological)
# Explanation of Evolution

THIS CHAPTER MAKES an auspicious claim in its title. As we saw in the preceding chapter, a shortcoming of the strictly scientific explanation is that it is not equipped to deal with the subjective dimension of experience. Thus, it is not prepared to address questions that people have that represent a more holistic human experience and understanding of evolution. As we shall also see, this shortcoming prevents empirical science from addressing the role of the future in understanding evolution. In this chapter, we will explore how a future-oriented way of understanding reality may offer a more comprehensive, intellectually satisfying view of the process of evolution than empirical science alone can offer.

## Why do scientific atheists and agnostics feel uncomfortable with Teilhard?

Daniel Dennett, an American philosopher of science, expressed his discomfort with Teilhard's notion of the subjectivity of the cosmos and his future-oriented thought in his book *Darwin's Dangerous Idea* by crudely caricaturing him to make him seem as though he was scientifically incompetent. He then adds, "The problem with Teilhard's vision is simple. He emphatically denied the fundamental idea that evolution is a mindless, purposeless, algorithmic process" (Haught 2000, 82). Perhaps by now, one may suspect in this statement a bit of empiricist dogmatism in that there is an implicit refusal to consider additional ways of understanding the evolutionary process besides scientific empiricism, as obviously basic and crucial as that method has been and remains in uncovering the empirically factual basis of evolution. It is interesting that Dennett used the word *fundamental* in the above-quoted statement. As hostile and unfair as Gould's attack on Teilhard's reputation was in his book *The Panda's Thumb*, in which Gould accused Teilhard of being involved in creating a scientific hoax, which has since been thoroughly debunked (Cela-Conde and Ayala 2007, 87; Costello 1981; Gardiner 2003; Gee 1996; Lukas 1981), Dennett's rude caricatures of Teilhard have been perhaps even more aggressively hostile. Teilhard's reputation as a scientist in his day was excellent (Haught 2000, 82; Huxley 1959; Morowitz 1997, 16–27). The prominent biologist Harold Morowitz, who doesn't accept Teilhard's theology, nonetheless claims to be an "admirer of Teilhard's biological writing" and that Teilhard was "insightful and ahead of his time." He recognizes that Teilhard anticipated punctuated equilibrium, the now well-supported, documented theory that evolution has occurred in a series of rapid bursts of change punctuated by long periods of relative stasis while those who take complete credit for it (Gould and Eldredge) were still "in short pants" (Haught 2000, 200).

Undoubtedly, many atheistic and agnostic scientists feel uncomfortable with passages in Teilhard's writings in which he integrates philosophical positions and religious faith statements with science. However, empiricism, the operating philosophy of science, is itself is a philosophical position. *Empiricism* asserts that knowledge arises from the sensory experience of the evidence (Markie 2008). The correspondence of this sensory experience with what is really there is assumed by empiricism. Thus, scientific empiricism itself is seen to rely on faith that sensory experience is a reliable index of what is real. Empiricism arises from *philosophical realism*, which asserts that reality is independent of our conceptual schemes. Realism holds that getting knowledge of the truth-value of a given concept is a process of continuing refinement of the concept to best reflect empirical observations of reality. So science involves a process of refining a concept so that it more accurately corresponds with what is observed. Science, then, not only places its faith in reality's actual correspondence to our minds' and senses' perceptions of what is observed (Miller 2008). Scientists must also trust that their processes of inductive reasoning correspond with reality when they build concepts from data and that the internal consistency shown by the systems of concepts that they build does in fact reflect reality. Haught puts it succinctly: attainment of knowledge is conditional on some kind of faith commitment. Faith commitment is not inevitably an obstacle to knowledge. In order to glean scientific information, scientists must be committed to the faith that the universe is intelligible and believe that truth about it is worth exploring (Haught 2000, 111).

JOSEPH FORTIER

# What is metaphysics, and how does it influence our understanding of evolution?

*Metaphysics* is another branch of philosophy that deals with explaining the fundamental reality of being and the world. Besides what may seem a conflation of faith with science to some in Teilhard's writings, his metaphysics of process toward deeper coherence and fulfillment in the future is in fact unscientific and thus may also be discomforting to some. *The Phenomenon of Man* is not intended to be a scientific treatise but rather to express Teilhard's sense of the integrity of a unified vision of truth that draws from both the understanding of empirical science and religious understanding regarding the evolutionary process. The future-oriented metaphysical angle Teilhard develops is an original and useful tool for formulating a more comprehensive theistic vision of evolution that incorporates scientific understanding, as we'll explore below.

Altogether, Teilhard's undertaking when he wrote *The Phenomenon of Man* was courageous. It certainly wasn't written to win any popularity contests. As we shall see, it can be seen as being both a point of articulation between science and a more comprehensive, philosophical view of evolution and as a hinge integrating the human experiences of knowing through science and understanding through faith. Had Teilhard been permitted by the Catholic Church and his religious order to publish in his lifetime, he would have been given the opportunity to address his critics to the benefit of his work (Haught 2000, 84).

*The nub of the issue: the problem of metaphysics of the eternal present*

A difficulty with which we in the West have contended, including scientists, is a metaphysics of the static, eternal present. Teilhard acknowledged that our religious thought has been dominated by this metaphysics of being, which was inherited from classical Greek philosophy and assimilated in some form by Christian, Jewish, and Islamic theology. It still dominates much of the thinking in the West. A central tenet of this metaphysics of the eternal present is that perfection and fulfillment already exist—a holdover from the Platonic notion of ideal reality of which all that we are and see is a vague, imperfect representation of what is most real and perfect. A medieval spinoff of this metaphysics was the notion of a hierarchical Great Chain of Being, with God at the top link of the chain and humans somewhere between the angels above us and other material creatures below us. This Great Chain of Being was considered to be static and eternally present (Haught 2000, 84–85). Attachment to such a view of reality would certainly pose difficulties for one attempting to understand and absorb the process-oriented view for which evolutionary science provides evidence. Such a view has also posed no little difficulty for resolving the theological problem of faith in a benevolent divine creator, given all the suffering and pain that occurs in the created order (along with the beauty and enjoyment). Somehow, it doesn't seem like perfection. Certainly, any philosophy or theology that takes all this evidence into account must embody another view rather than such a metaphysics of the eternal present. Such a view would be much more akin to that of Teilhard's, as we shall see.

## Is the metaphysics of the past implicit in evolutionary science a satisfactory solution?

The method that science uses for investigating natural history and uncovering evidence for evolution necessarily looks to the past for its evidence. Fossils are analyzed, dated, and compared. Organisms still living today are compared using morphological and biochemical methods to elucidate their phylogenetic relationships; that is, how closely or distantly related they are with respect to past common ancestry. The principle of uniformitarianism was invented by the medieval Persian intellectual Avicenna, modernized by the Scottish geologist Thomas Hutton, and popularized by another Scottish geologist, Charles Lyell. Uniformitarianism assumes that natural physical processes operating in the universe now have always operated in the universe in the past, that the rates at which these processes operate have not changed, and that the same laws of physics apply everywhere in the universe. When applied to the behavior of matter and energy at the chemical and biological levels of organization, the evidence consistently shows that uniformitarianism is trustworthy and informative, yielding evidence that is congruent with other kinds of evidence for evolution. Altogether, the empirical evidence for evolution, in the context of the uniformitarian view of natural history, represents a view oriented toward the past for answers to our questions concerning the fundamental reality of being and the world. The scientific method has

JOSEPH FORTIER

indeed uncovered comprehensive, incontestable evidence for evolution. Because past-oriented evidence works so well for empirical investigation of evolution, the materialist reading of evolution might be called a "metaphysics of the past." Haught notes that such a materialistically based metaphysics places the source of life's diversity and complexification in a purely physical determinism. This past-based, deterministic scenario has led—step by accidental, contingent step—out of the nonliving past and into the present in all its profuse complexity and diversity of life. As we saw (chapter 6), Kauffman finds evidence for the emergence of novelty that is probably based on primordial, universal, as yet not demarcated laws rather than on accident and contingency. Interestingly, in doing so, Kauffman takes a step away from past-based metaphysics and a step toward a metaphysics of the future. He addresses the issue of spontaneously emerging complex novelty in the future, not strictly reducible to laws governing its past-based components. Nevertheless, he sees this process of emerging complexity as being driven by "mysterious laws" that originated in the past that determine that new "order for free" condenses out of complex, networked systems in which there is spontaneity at the edge of chaos. Haught asks whether such a past-oriented deterministic view really allows for the emergence of true novelty (Haught 2000, 86). Perhaps any notion of true novelty in the impersonal, deterministic universe envisioned by a purely empiricist-based metaphysics of the past is a banality rather than *true* novelty.

*Teilhard revisited: what about future-oriented metaphysics?*

Perhaps part of the reason some scientists and philosophers of science, such as Gould and Dennett, feel uncomfortable with Teilhard's dangerous idea is not only because of his giant step toward an evolutionary theology of nature that integrates natural science with concepts and ideas of philosophy and theology, but also because of his giant step into a metaphysics of the future. Scientific investigation, after all, had led us out of the stagnation of the eternal present, which threatened to stifle our imaginations and frustrate our acceptance and appreciation of the "grandeur in this view of life." This was accomplished by replacing a metaphysics of the eternal present with an implicit metaphysics of the past. For Haught, metaphysics of the past is certainly not identical with science but is rather a consequence of how scientists define what is real and what is not real. Scientists confine their definition of reality to what is empirically knowable (Haught 2000, 83), namely, to what is observable and measureable to the senses. Thus, scientists' view of what is real is constricted by an implicit, materialistic metaphysics of the past since the past is knowable by empirically observable evidence (chapter 4) and the future is not. But such a materialistic, deterministic metaphysics doesn't really address certain vital questions that people have, such as why there is such an overarching tendency for matter to evolve toward life, mind, and spirit. Unfortunately, Haught does not always make a clear distinction between scientists who are persons and the *subset* of those persons who are scientists who advocate empirical science as the *only* way to know anything. Thus, in places, he seems to suggest an unnecessary dichotomy between scientists and theologians. This certainly does not describe Teilhard and those scientists who find areas in which to agree with him, such as Julian Huxley, John Polkinghorne, and Arthur Peacocke, as Haught would readily admit.

Teilhard explores a metaphysics of the future because such an angle on the fundamental reality of being and the world allows for a deeper, broader, more satisfying explanation of the data of evolution, such as the emergent novelty of life, mind, and spirit. It is important to make clear that a metaphysics of the future fills out and completes an understanding of the data presented by a materialistic metaphysics of the past, rather than replacing it. With respect to an evolutionary theology of nature grounded in the evidence of empirical science, it's obvious that there can be no authentic metaphysics of the future without the empirical data that is associated with a metaphysics of the past. Just as materialistic empiricism is the philosophy most practical for systematically working out the tremendous body of knowledge about the universe that science renders, our understanding of the universe is perhaps most effectively completed and filled out with theological investigation. But now, what does such a metaphysics look like? Although we've glimpsed at Teilhard's pioneering of this metaphysics in chapter 7, especially with respect to Omega, John Haught, Senior Fellow of Science and Religion at the Woodstock Theological Center in Georgetown University and expert witness on behalf of Kitzmiller et al. in the *Kitzmiller et al. v. Dover Area School District* case (chapter 8), is the premier navigator of this investigation to date.

*Theology assumes God. How does God in any way*
*contribute to an intellectually satisfactory position or solution?*

Theologies of nature a priori assume a given theistic tradition and then proceed to explore how the findings of science enlarge the tradition's

JOSEPH FORTIER

understanding of God and God's relationship with the world (chapter 5). Before we explore Haught's Christian evolutionary theology of nature, it is fitting in such a book as this to address why belief in God might be an intellectually satisfactory position to take in the first place. Here I appeal to the thought of the Templeton Prize–winning British Anglican theologian and doctor of biochemistry Arthur Peacocke.

Peacocke notes that the existence of the world, or "all that is," is not self-explanatory. If one asks, "Why is there anything at all?" science cannot provide an answer. No scientific account is possible for the fact of the existence of whatever it was from which the universe emerged and expanded 15 billion or so years ago. No scientific account is possible for the existence of the relationships that scientific investigation perceives and measures within this universe. In addition, the existences of neither the universe nor these relationships within it are logical necessities. Thus there neither is, nor can be, any scientific account for the very existence of a universe such as the one of which we are aware or of any other universe (Peacocke 2001, 39).

The best explanation that can be inferred, Peacocke holds, for the existence of "all that is" and the fundamental physical laws governing it, is that this "all that is," with all its processes of beginnings and becomings, is grounded in another reality. This other reality must be the source of the actual existence of "all that is" and, thus, a singularity. This is so because the existence of the universe that we know, and those of any other possible universes, is contingent: without a scientific account and without logical necessity. The source of this universe's existence must be self-existent, the only reality that is itself the source of its own being, not contingent and, thus, the ultimate ground of being and a singularity. Thus, the mystery of existence points to an ultimate reality that somehow is the source of the existence of "all that is" (Peacocke 2001, 39–40). Peacocke notes the inadequacy of words to circumscribe the ultimacy and transcendence of this self-subsistent reality. Given this reality's transcendence, it is inexpressible by us and can be referred to only by metaphor, analogy, and assumptions based on extrapolation (Peacocke 2001, 40).

The ultimate reality that is the source of all contingent being, according to the best explanation that can be inferred, must be a unity. This is so for two reasons. First, if the putative ultimate reality were multiple, that is, separate realities, then the question of the origin of this multiplicity would be begged. Thus, multiple ultimate realities would not really be ultimate. Second, the universe discovered by empirical science is itself one interlocking network of various kinds of entities, all interrelated by the same regularities and laws: one world. Once again, a multiplicity of ultimate realities would beg explanation as such a precondition for the universe that empirical science reveals, with its interlocking, relational unity, and would be unfeasible and incoherent (Peacocke 2001, 40). Thus, the ultimate reality in which the universe is grounded and finds its source must be a unity.

Besides being a unity, the universe, including the earth, reveals awesome diversity in its constituent entities, structures, and processes, including its diversity of multiple levels of complexity. Thus, the ultimate reality must be such that it has the ability to give existence to this wide diversity of entities. The best explanation that can be inferred from this is that the unity of the ultimate reality must not just be mere simplicity but must rather be a diversity in unity, a being of awesome richness itself, capable of rich, complex depths of expression (Peacocke 2001, 40).

The relationality of the world, in a similar way, suggests a relational ultimate reality. Out of this matrix of relationality of the world, humans emerged, with the distinctiveness we recognize in our own and others' personhood. Each human entity, by virtue of his or her capacity for social interaction and the freedom and consequent ethical nature and capacity for freely chosen love that comes with self-reflective consciousness, becomes a uniquely endowed person. The best explanation that can be inferred from a world in which communities of unique persons and personalities emerge is that the ultimate reality, also, must be at least capable of personhood and personality. Thus, this ultimate reality is more properly designated as he/she rather than "it" since personal predicates are appropriate for persons. The Ultimate Reality is at least a person, since again, language fails in the task of circumscribing the transcendent Ultimate Reality that gives existence to "all that is." Peacocke identifies the ultimate existent ground of being as God in the English language, with various cognates of this same idea in other languages (Peacocke 2001, 42–43).

In summary, the best explanation that can be inferred for the existence of "all that is," or the universe, is that it is grounded in another ultimate reality. This is so since the existence of the universe, being neither accountable by science nor logically necessary, is contingent. Given qualities we observe in the world (the universe) such as its fundamental unity yet awesome diversity, the Ultimate Reality must itself be fundamentally one yet diverse. This is so especially since any multiple Ultimate Realities responsible for a reality that manifests itself as a unity would beg explanation and thus would itself be contingent and because such a giver of existence to the universe, with its awesome diversity in unity, must be a being of awesome richness itself, capable of rich, complex depths of expression. Finally, the Ultimate Reality of a universe from which emerge conscious, deliberative, self-aware, morally responsible persons capable of freely deciding to love must him/herself be at least a person with at least these qualities of personhood. The above qualities of a noncontingent Ultimate Reality follow from inferring the best explanation for the existence of a contingent universe with the qualities that the universe manifests.

*Is a theology of nature that integrates God, evolution, and*
*future-oriented metaphysics an intellectually satisfactory solution?*

For Haught, the novel informational possibilities that emerge in the evolutionary process do not arise from the grinding onward of the algorithms of the past but rather with the arrival of the future. The novel informational possibilities made available to the evolutionary process are presented by the constantly dawning future. The unpredictable, spontaneous novelty of the "order for free" that Kauffman describes as condensing out of highly complex systems at the edge of chaos (chapter 6) after a threshold of complexity has been crossed is, for Haught, a fruit of the future dawning on the present, full of new informational possibilities from which to choose. Evolution, change over time, is only possible because the future faithfully presents relevant new possibilities as it continually opens. A contingent event is one that is not certain at a present moment in time since it depends on another event that itself isn't yet a certainty. Contingent events are neither logically necessary nor impossible. They may only be verified by sensory observation. They may happen by chance, without a perceivable cause. Contingent events are sometimes described as "fortuitous" or "accidental." An example is the shuffling of genes between chromosomes, as in a crossing-over event during meiosis, the process by which gametes (egg and sperm cells) are formed (chapter 2, figure 2.9). Whether two similar chromosomes "decide" to exchange genetic material or not and what genetic material they "decide" to exchange are fortuitous and accidental factors of a random event. This and other contingent, coincidental, random events that S. J. Gould describes in his work don't ultimately explain the novelty that continually emerges in evolution, although this language of chance does explain the lack of determinism involved in the process. The occurrence of these events, Haught notes, are dependent more ultimately on time's opening into the future to provide temporal space for these events to play themselves out. The ultimacy of time as an explanation for evolution becomes clearer when we consider the relationship between time and contingency. The occurrence of contingent events, such as crossing over during meiosis, doesn't bring about the future. Rather, the arrival of the future provides the temporal space that allows events to relate to each other as contingent, that is, to not just be inevitable outcomes of past deterministic causes (Haught 2000, 87).

A metaphysics of the future is expressed in the biblical tradition that finds the abode of ultimate reality to be limited to neither the causal past nor some fixed, timeless present "up above" but rather found in the constantly arriving future. The advantage of this view of reality, although strange to us, is that it accommodates both the data of evolutionary science and biblical claims about the way a god of promise relates to a world in process of becoming, longing and groaning toward its fulfillment (Haught 2000, 88; Paul, letter to Romans 8:18–25). For example, a shortcoming of a purely materialistic view, associated with a past-oriented metaphysics of the evolutionary process (however valuable its ability to bring to light hard empirical evidence for this process), is that it is unable to take the fact of subjectivity into account and is thus unable to fully address emergent novelty. Thus, it can provide no complete illumination with respect to the emergence of what we all experience as our subjectivity: our peculiar, individual ways of reacting interiorly to the world and one another, just as it is unable to fully address Teilhard's "within" and interior "resonance to the All." To describe the human subjective capacity for universal love, Teilhard writes, "How can we account for that irresistible instinct in our hearts which leads us towards unity whenever and in whatever direction our passions are stirred? A sense of the universe, a sense of the *all*, the nostalgia which seizes us when confronted by nature, beauty, music—these seem to be an expectation and awareness of a Great Presence . . . Resonance to the All—the keynote of pure poetry and pure religion" (Teilhard de Chardin 1959, 266). For Haught, it is this ongoing process of intensification of the subjective "within" or "inwardness" that he calls "novelty" entering the universe (Haught 2000, 88) and to which he feels materialism is closed a priori from recognizing and taking into account.

As the reader has probably guessed, a metaphysics of the future is a religious idea. Interestingly, the award-winning scientist and agnostic Kauffman also found the need to address the future in order to more

JOSEPH FORTIER

fully explain the evolutionary process and wrote about his sense of the spiritual in the universe. Metaphysics of the future is deeply rooted in the subjective experience that people have of something that to them is powerfully and incontestably real: what may be called "the power of the future." Only by placing oneself in a religious attitude of hope can one find the opening to the experience of this power. Haught says that in the biblical context of faith, this power is the subjective experience of being grasped by "that which is to come." To reiterate Teilhard, "A sense of the universe, a sense of the *all*, the nostalgia which *seizes us* [italics mine] these seem to be an *expectation and awareness of a Great Presence* [italics mine] . . . the keynote of pure poetry and pure religion" (Teilhard de Chardin 1959, 266). Paul Tillich, a great twentieth-century Protestant theologian, describes this experience of the future as sense of being grasped by the "coming order." He writes, "The coming order is always coming, shaking this order, fighting with it, conquering it and conquered by it. The coming order is always at hand. But one can never say, 'It is here! It is there!' One can never grasp it. But one can be grasped by it" (Haught 2000, 89). For the twentieth century Jesuit theologian Karl Rahner, as for Teilhard, this future constantly meets us at the nexus of the present passing into it. This future that comes to meet us, takes us, and makes us new. In this sense, in what Rahner calls the future's absolute depth and what Teilhard called "Omega," the future may be experienced as the effectiveness of God, a manifestation of the Taoist *wu wei*, or noninterfering effectiveness, that creates space in which creation can exist and continue to become the process that is itself. For Haught, this efficacious future is simply God (Haught 2000, 90). It would seem to follow, using Teilhard's notions of the subjective "within" of all things and of reflective human consciousness as a product of the effect of the evolutionary process on this "within" in the human lineage, that this conscious human experience of the future would also pertain to the relationship with the future of all entities in the universe. As Haught puts it, "being grasped" by the absolute future would seem not to just pertain to ourselves but to the entire cosmic process. In other words, the entire universe is constantly drawn forward by a divinely renewing future. In this understanding of the power of the future, an intuition that is religious in origin, all things receive their existence from an endlessly bountiful future that Haught calls God (Haught 2000, 90). It can be seen that this understanding of divine creativity involves the notion that the past and the things that existed in the past and the present, have been and are in some way given their status of existence by the always arriving but always just ahead future. Only this understanding of creativity satisfactorily explains the *degree* and *quality* of novelty that has arisen in the world, including life, mind, and spirit, subjective as well as objective experience. Thus, the ultimate source from which arise new life, new species, and surprising innovation based on new arrangements of complex organization, is only to be found in the future (Haught 2000, 90–91). Only such an ultimate source provides explanatory power for a contingent universe as its ground. Only such an ultimate source provides satisfactory explanatory power for the emergence of such diverse, mysterious novelty as the degree, quality, and power of subjective experience as exists in human hearts and souls, having evolved out of the primordial cosmos. An ultimate explanation for all that happens in cosmic and biological evolution as being the unfolding of what has happened before, with the sole mechanism of complexification lying in some mysterious, presumably materialistic law of an admittedly contingent universe, does not answer basic questions. Why is there a contingent universe at all? Why do we experience our subjectivity as we do, as powerfully as we do? Why the strong movement in us elicited by nature, music, art, poetry, religious meaning, that is experienced as transcendent mystery? Why the hunger, thirst for ultimate value? Is it really to be answered by genetic mechanism and neurology alone, using only the past-oriented materialistic way of explaining, finding these answers presumably at some time in the future? If genetic or neurological mechanisms are found for these deep, ultimate subjective experiences, as they likely will be, at least in part, are we really to believe that such a past-based mechanical explanation is, by itself, an *ultimate* explanation for the deep sense of meaning we experience in our subjectivity? Why the assumption of the new scientific atheism that only objectively derived understanding is real and true while our deep, subjectively derived understanding of ultimate beauty, being, and value (religious experience) is dismissed as superstitious delusion? Is it that we are complacent, secured in a cave, secured in a shadowland of the past, playing on the cave walls that are but incomplete reflections of the bright light spilling into the cave from the opening? Is it possible that while of course there is invaluable information to be gleaned from those cave walls, nonetheless that information beckons us to summon the courage to also turn and walk boldly into the bright light of the future, the source of the events played out on those mossy walls, plus so much more beauty, meaning, and value?

JOSEPH FORTIER

What specifically is the contribution to an
evolutionary theology of nature that Christianity brings?

It is always possible to come up with a description of God that will be
compatible with whatever one wants God to be compatible with—such as,
for instance, empirical science. But of course, such a tactic lacks credibility,
as it can lead to conjuring an image of God that is vague and religiously
uninformed. It is more honest, and thus more credible, to ask whether the
understanding of God in a concrete religious tradition is consonant with
contemporary scientific understanding. Christianity understands God as
the god revealed by Jesus: compassionate, vulnerable but strong, a personal,
relational God of suffering love. For Christian theology then, according
to Haught, the task is to seek to understand the natural world, with its
evolutionary history and character, in terms of hope in the outpouring of
God's compassion associated with the God of Jesus: the crucified and risen
one (Haught 2000, 110).

In chapter 8, we saw that divine power can be seen as God's participation
in the world's suffering since authentic love cannot compromise the
dignity of the beloved by coercing his/her decision making and process
of becoming, such as to make the beloved behave in such a way that (s)he
won't suffer. In this sense, God's humility and suffering love implies a sort
of defenseless vulnerability in God's solidarity with the ongoing process
of existence, which can powerfully and effectively disarm evil, rather than
weakness and powerlessness. Mohandas Gandhi and Martin Luther King
Jr. displayed this sort of power in the twentieth century. In doing so, both
found their inspiration in the life of Jesus (chapter 8). This notion of divine
omnipotence expressing itself in the world as defenseless vulnerability is
alien to human notions of power, with their capacities for destructiveness
and coercive control. The Christian experience of God in the life of
Jesus is the experience of the self-emptying (*kenosis*) of God (Philippians
2:5–11) that leads to promise of renewal and growth of relationship
with God. It is the discovery of something new about God and God's
power: that God's power is self-limiting and humble and is more about
respectfulness and nurturing of the dignity of that which emerges from
its creativity than about commanding, controlling, and punishing. From
this specifically Christian angle, Haught as Christian theologian, reflecting
on the relationship of evolutionary science to religion, sees God as both
kenotic, self-emptying love and creatively persuasive power of the future.
This sense of God as self-humbling love that opens up a new future for

the world emerged in Christian awareness specifically with respect to the Christ event (Paschal Mystery): the suffering, death, and rising from the dead of Jesus, which is at the core of Christian faith. From this awareness, a future-oriented theological understanding emerges, which takes into account both evolutionary science and the portrait of a vulnerable god who is faithful to his relationship with contingent existence in process and thus powerfully, but not coercively, effective. With this understanding, nature's evolutionary journey may be seen as showing the topography of subjective, relational meaning that wouldn't stand out without a prior commitment to such a way of understanding.

What exactly is meant by "power" in the above sense? In contrast to coercive power, the power in the vulnerability of a man killed on an instrument of torture out of divine love for the world is the power to influence the world. Haught holds that this manner of expression of divine power ultimately explains the chaotic randomness, struggle, and apparent meandering disclosed by the evolutionary account of the natural history of life as the underside of incredible creativity on the part of the world in process. This underside of apparently meaningless meandering and suffering is due to the lack of coerciveness on the part of an authentically loving God, willing to suffer with the odyssey of the evolutionary process of that which is loved. On the other hand, the complex pattern and beauty that spontaneously emerges is also consistent with the notion that the "universe as it is" is the consequence of infinite love. In God's desire for intimacy with the universe, God insists on preserving the unique journey of this beloved world toward becoming itself, a being and becoming that depends on its own authentic otherness from God (Haught 2000, 113–114). Only in this way can authentic relational intimacy be attained, in contrast to the alienation inherent in any relationship defined by manipulation or coercive controlling.

With respect to human freedom, Haught holds that this can only emerge in a world that is independent from God, such as the one we know. Impersonal laws such as gravity, natural selection, and self-organization are manifestations of the independence of the universe from God. So the conception of God in Haught's evolutionary theology of nature is the antithesis of the Deist conception of God in which God controls the universe through such mechanisms. This noninterfering self-distancing of God though is an intimate expression of God's involvement in the world as an effective presence seeking authentically loving relationship with an authentic other, in which dialogic intimacy is possible. The deep autonomy that emerges in this world because of God's self-distancing love appears most obviously in the emergence, by the process of evolution, of human freedom (Haught 2000, 114).

Jesus is the ultimate expression of God's love for the world. God's solidarity and suffering love for an autonomous world, whose autonomy God values and maintains, is graphically expressed in Jesus's life, suffering death, and resurrection from the dead as a consequence of his faithfulness to his divinity, faithful love for this world, and faithfulness to maintaining his authentic, noncontrolling relationship with this world. The consequences of the continuing relationship with the God of Jesus after the Jesus event are manifesting themselves and will continue to manifest themselves. God's love is manifested in humans to the extent that humans continue the long process of opening their collective awareness to the Jesus event and its timeless, ultimate significance—the ultimate significance for human dignity and that of the earth and universe and the ultimate significance of God's immense love for that dignity of being.

Out in the forest there are two clear-water streams that, at close range, apparently flow from different mountaintops but which when viewed from farther away, can be seen to flow from two ridges of the same mountain. Somewhere ahead in the forest, the two streams join and become confluent.

# Summary

☐ Daniel Dennet, Richard Dawkins, and others have found difficulty and expressed hostility with Teilhard de Chardin. This is in part due to an uninformed literal interpretation of the philosophy of scientific empiricism when applied to evolution, which impedes the open-mindedness necessary to read Teilhard in context. It is in part due to Teilhard's association with faith statements occurring alongside scientific thinking in his work and also to his future oriented metaphysics.

☐ *Metaphysics* is branch of philosophy that deals with explaining the fundamental reality of being and the world. Scientific investigation of evolution has, to great advantage and benefit for humankind, employed a *past-oriented metaphysics*, which explains evolution in terms of solid empirical evidence based on past events. However, scientific investigators who assume that past-oriented metaphysics is the *only* explanation for the fundamental reality of being and the world step past their competency and err. While empirical science is powerfully competent in discovering objective factual data, subjective data is ultimately beyond its competence.

☐ A contribution that the scientific investigation of evolution has made for human thought is its challenge to a traditionally assumed *metaphysics of the eternal present*, which at least in the West has originated in Greek thought, was adopted by medieval Christian thought and has had the effects of (1) frustrating Christianity's understanding of the value of evolutionary science and (2) created the "theodicy problem" of squaring a good, eternally unchanging, all-powerful, controlling God with evil and suffering in the world.

☐ Teilhard de Chardin and Haught propose a *metaphysics of the future*, which understands the value of the process thought of evolutionary science and also understands God's activity as presented from the future. This understanding of God's activity sees God's power as inviting, attracting, and persuading the dynamic and relational evolutionary process of the universe into the space for it that is continually created as the present moves into the future toward God.

☐ The question arises, why is there anything at all? The universe is contingent, coming into existence about 15 billion years ago.

The idea of one ultimate, self-subsistent god who is a person is an intellectually satisfactory position in that it offers a best explanation that can be inferred for the existence of a contingent universe (or universes). The contingent universe shows itself as consisting of a wide array of diverse entities with a wide, rich array of interrelationships. In addition, human persons with reflective awareness, freedom, and thus ethical capacity beg explanation. The best explanation that can be inferred for the existence of a universe with these qualities is an ultimate reality that is relational, at least a person, and a unity of awesome richness capable of complex depth of expression.

☐ Theologies of nature assume the truth-value of their traditions and seek greater accuracy in how those traditions are expressed and interpreted in light of updated scientific findings. Haught's evolutionary theology of nature begins with the discoveries of evolutionary science and Teilhard de Chardin's insights regarding a future-oriented metaphysics as a method for understanding the fundamental reality of being and the world. A theology that incorporates a metaphysics of the future understands God's relationship and activity as located in the future, to which we are constantly drawn. God is not seen as controlling by power but rather whose power is respectful and loving and, thus, manifests itself as persuasive and attractive.

☐ For Christians, Jesus is the manifestation of divine power, precisely in the self-emptying of God that is seen in Jesus's suffering, death, and resurrection. God's relational love for the world and humans in humbling himself to become in solidarity with the material universe, the ecology of the earth, and the human condition in the person of Jesus as a suffering servant is a powerful, creative act in a humble, persuasive, attractive, noninterfering way. As this reality steeps in the universe, all things are potentially drawn by this love toward God, and human awareness undergoes a process of awakening to this love and may be drawn into a dynamic, living relationship with this noninterferingly, effectively persuasive God whose activity is love. The latter, however, can only happen to the extent that persons give themselves to this process in acts of trust.

# References

Cela-Conde C. J. and F. J. Ayala. 2007. *Human Evolution*. New York: Oxford University Press USA.

Costello P. 1981. "Teilhard and the Piltdown Hoax." *Antiquity* LDC: 167-171.

Gardiner B. 2003. "The Piltdown forgery: a re-statement of the case against Hinton." *Zoological Journal of the Linnean Society* 139: 315-335

Gee H. 1996. "Box of Bones 'Clinches' Identity of Piltdown Paleontology Hoaxer. *Nature* 381: 261-262.

Haught J. 2000. *God After Darwin: A Theology of Evolution*. Boulder, Colorado USA: Westview Press.

Huxley J. 1959. "Introduction." In *The Phenomenon of Man*, translated by Bernard Wall, New York: Harper & Row, Inc.

Lukas M. 1981. "Teilhard and the Piltdown 'Hoax'." *America*, May.

Morowitz H. J. 1997. *The Kindly Mr. Guillotin and other Essays on Science and Life*. Washington D.C.: Counterpoint.

Paul. 1977. "Letter of Paul to the Philippians." In *The New Oxford Annotated Bible with the Apocrypha*, May H. G. and B. M. Metzger (eds.). New York: Oxford University Press

Paul. 1977. "Letter of Paul to the Romans." In *The New Oxford Annotated Bible with the Apocrypha*, May H. G. and B. M. Metzger (eds.). New York: Oxford University Press

Peacocke A. 2001. *Paths from Science towards God*. Oxford, Great Britain: Oneworld Publications.

# Glossary of Theological and Philosophical Terminology

**anthropic principle**. A development in natural theology which takes into account that astrophysicists realize that the universe seems fine-tuned for the emergence of life, including human life.

**argument from design**. A method of doing natural theology that was developed by the Anglican cleric William Paley, in which one could prove the existence of God from the evidence of intricate design in nature, such as that of the eye.

**biblical literalism, ahistorical**. Interpretation of a given biblical passage based on how the words in the passage have come to be understood in the modern context rather than taking into consideration evidence from modern scholarship as to how they were understood when they were written.

**biblical literalism, informed**. Interpretation of a given biblical passage based on the intention of the biblical author's use of words.

**Christian Fundamentalism**. A movement that began in the United States in the first decade of the twentieth century and ascribes to a fourteen-point creed established at the Niagara Bible Conference (1878–1897). The first of these points states that the fundamentalists believe that the Holy Ghost gave the very words of the sacred writings to holy men of old.

**creation science**. A religious movement, based in Christian fundamentalism, which seeks to use evidence used by evolutionary scientists to instead show that an ahistorical biblical literalist interpretation of how the universe came about is the better explanation.

**deism**. A philosophical position that understands God as the architect-designer of the universe and its laws, who then steps back from any relationship with it to let it take off on its own.

**empiricism**. A theory of knowledge that knowledge is dependent on sense experience.

**intelligent design**. A religious movement, based in Christian fundamentalism, which holds that complex biological mechanisms such as the eye are "irreducibly complex" and thus could not have evolved. While intelligent design attempts to pose as science, it was unmasked as such in the *Kitzmiller et al. v. Dover Area School District* trial in Pennsylvania, USA, in 2005.

**kenosis**. The self-emptying, suffering love shown by God in Jesus's life that offers hope for growth in relationship with God.

**logos**. An originally Greek idea that was used by the Hellenized Jewish thinker Philo to mean "creative principle" or "wisdom of God." The word came to be used by Christians to refer to Jesus as the creative Word of God.

**metaphysics**. A branch of philosophy that deals with explaining the fundamental reality of being and the world.

**monogenism**. From early twentieth-century Catholic theology, a position that maintains that the human species originated from only two original human beings.

**natural theology**. A way that thinkers have attempted to find an integration between science and religion by looking to science for evidence of the existence and activity of God.

**objective**. Adjective referring to what is outwardly observable in a person or object.

**panenthism**. A philosophical position that understands God as transcendent to the universe yet as intimately present in the universe.

**pantheism**. A philosophical position that identifies God with the universe.

**philosophical realism**. A philosophical position that holds that reality is independent of our conceptual schemes and that during any given time period, how we see things is only an estimation of reality to which we approximate more closely by the process of learning from observation.

**process thought**. Developments in philosophy and theology that view reality as an ongoing process, similar to the evolutionary process.

**reductionism**. A philosophical position that asserts that the laws and theories of all sciences including biology, psychology, and sociology are reducible to chemical and physical laws, and that the physical and chemical components of any complex system, such as living systems, ultimately determines its behavior.

**scientific materialism**. A philosophical position that asserts that the most fundamental reality in the universe is matter and that the only way to know anything is by the scientific method.

**subjective**. Adjective referring to what is inward and not easily observable in a person or object.

**Tao**. In the Chinese philosophy/religion Taoism, it is the ultimate reality, which is energetically passive but informationally active. **w*u* *wei***. The way that Tao works in the world, which is a way of noninterfering effectiveness (chapter 5).

**theodicy**. A branch of philosophy and theology that explores the problem involved in the simultaneous existence of evil, and goodness of God.

**theology of nature**. A way of seeking an integration between science and religion in which the fundamental insights of a religious tradition are assumed, as well as the insight that there is a unified truth that the human mind can approach, and science is investigated in order to better understand and articulate the religious tradition in light of scientific findings.

# INDEX

# D

Darwin, Charles, 34, 56, 97, 164
  and observation on
    mockingbirds, 35
  *On the Origin of Species*, 37, 88,
    103, 118, 139, 155
Darwinian evolution, 221, 225,
  229, 231
*Darwin's Dangerous Idea* (Dennett),
  233
daughter strand, 74
Dawkins, Richard, 226, 248
deism, 148
Dennett, Daniel, 233–34
  *Darwin's Dangerous Idea*, 233
deoxyribose, 63
descent with modification, 35,
  54–55, 120
Devonian period, 109
diabetes, type 1, 112
dialogue, 137, 148, 159
dignity of humanity, 141, 159,
  199, 247
dinosaurs, 170
diploid, 39–40
*Dll* (*Distal-less*) gene, 115
DNA (deoxyribonucleic acid),
  63–64, 71
  base sequence, 83, 112
  polymerase, 74, 153
  replication, 74
dominance, complete, 42
dominant genes, 36, 39, 45, 58
*Dorudon*, 109
*Drosophila melanogaster. See* fruit
  flies

# E

Eden, 27
edge of chaos, 171–72, 215
*edu*, 27
egg cell. *See* gametes
Egypt, culture of, 21, 24
Einstein, Albert, 150, 156
Eldredge, Niles, 166
electron microscope, 67
Elohim, 19
embryo, 118
emergence of order, 166
empiricism, 33–34, 138, 142,
  234
*Endless Forms Most Beautiful*
  (Carroll), 114
Enkidu, 27–28
Enlightenment, Age of, 33, 164
*Enuma elish*, 19–21, 23–24
*Equus domesticus*, 105
*Escherichia coli*, 69, 112, 127,
  180
Ettinghausen, Andreas von, 39
Eve (biblical first woman), 29, 31
  as first ancestor, 17, 25
  meaning of the name, 29
evo-devo, 113, 127
evolution. *See* descent with
  modification
evolutionary theology, 221–22,
  224, 230
*Experiments on Plant Hybridization*
  (Mendel), 37
extinction spasms, 170
eyeless gene, 113, 127

Human Genome Project, 116
*Humani Generis*, 140, 162
human inflorescence, 200
Humulin, 112
Hutton, Thomas, 236
Huxley, Julian, 201
hybrid zone, 121
hydrogen, 180
hypotheses, 34, 56
*Hyracotherium*, 105

## I

*Ichthyostega*, 109
igneous rock, 99
Ijdo, J. W., 117
illusion, 149
inbreeding depression, 45
Incarnation, 210–11
independence, 137, 146, 159
   religious standpoint on, 147–48
   scientific standpoint on, 146–47
independent assortment, law of, 47
information, 151, 154, 160, 244
   according to Barbour, 152
   according to Haught, 151, 153
   according to Polkinghorne, 153
   in biological systems, 152
insulin, 112, 130
integration, 137, 154, 159–60, 253
intelligent designer, 141, 221
intelligent design theory, 155, 160, 221–22, 230
isotopes, 101, 131
Israel
   capture of, 19, 22, 30
   keeping records and traditions of, 22

## J

Jacobson's organ, 111
Jesus, 147, 210, 226, 245
   divinity of, 226
   as Jewish Messiah, 210
   as one high priest, 208
   as Son of God, 211
   as ultimate expression of love, 247
Jewish
   lifeway, 136
   theology, 24
John Paul II (pope), 157
   affirming evolution, 140, 200
   on Galileo, 139
Johnson, Phillip, 226

## K

karyology, 117
Kauffman, Stuart, 143–44, 159, 166, 182, 192, 201
   *At Home in the Universe*, 164
   on cells, 195
   on creation science, 165
   defining free will, 144
   on dialogue, 145
   *Reinventing the Sacred*, 143, 174
   on the universe, 180–81
Kelvin, Lord, 100, 103
kenosis, 245
keratin, 106
kilobase, 114
King, Martin Luther, Jr., 144, 209, 245
Knott, Max, 67
*kohen*, 208, 219

polymer, 64
polysaccharide, 64
Pontifical Academy of Sciences, 158
priest, 207–8
priestly author, 19, 30. *See also*
    creation, priestly and Yahwist
    narrative of
proteins, 49, 67, 69, 79–81
  enzyme, 79
  structural, 79
psychogenesis, 197–98
psychosphere, 216
punctuated equilibrium, theory of,
    166, 195
Punnett, R. C., 40
purines, 71
pyrimidines, 71
pythons, 110

## Q

quantum theory, 173

## R

radial energy, 192, 212, 215
radioactive decay, 189
radiometric dating, 101
radium, 189
Rahner, Karl, 243
recessive genes, 39, 45
red blood cells, 86
reductionism, 142, 159, 173
reflection, 198
*Reinventing the Sacred* (Kauffman),
    143, 174
religious institutions, influence of,
    144

religious thinkers, 142
RNA (ribonucleic acid), 67,
    80–81, 113
  polymerase, 81
*Rock of Ages* (Gould), 146, 162
R-strain, 65, 67
Ruska, Ernst, 67, 123

## S

salamanders, 122–23
Schillebeeckx, Edward, 226
Science of Complexity, 144, 158,
    166, 171, 181
*Science on Trial* (Futuyma), 145,
    162
scientific materialism, 137–38,
    142–43, 188, 253
scientific materialists, 138, 142
scientific method, 97, 138, 236
scientism. *See* scientific materialism
sea squirts, 114
sedimentary rock, 98–101, 126,
    131–32
segregation, law of, 47
self-organized criticality, 171
seven, 23, 25, 31
shale, 99
Shamhat (temple prostitute),
    27–28
Shapiro, Robert, 179–80, 182
  *Origins*, 179
sickle-cell anemia, 85
Smith, William, 100, 103
*Sociobiology* (Wilson), 142, 146
speciation, 98, 120, 125
sperm cell. *See* gametes
S-strain, 64–65, 67

stamen, 37
Stoeger, William, 147–48
*Streptococcus pneumoniae*, 64
Sturtevant, A. H., 44, 83
Sumerian civilization, 20, 27
supernatant, 69
Swinburne, Richard, 155
system
   closed, 151
   open, 172, 175, 184

## T

T2 phage, 69
tangential energy, 190, 215
Tao, 153, 228, 239–42, 249, 253
Taoism, 153, 228–29
*Tao Te Ching*, 228
Teilhard de Chardin, Pierre, 140,
   143, 186–87, 193, 199–201,
   214–16, 225, 242, 248
   on complexification, 191, 194
   on metaphysics of the future, 238
   on metaphysics of the present,
   235
   *Phenomenon of Man, The*,
   187–88, 193, 195, 235
   and the universe, 190
   on the universe, 188
   and "within," 188, 196
*Teilhardina magnoliana*, 197
telomeres, 118
Tetrateuch, 19
theology, natural, 154–56, 160
theory, 34
Thomas Aquinas, 147
thymine, 71–74, 80, 85, 91, 93–94
Tillich, Paul, 243

Torrance, Thomas, 150
toy model, 178
transcription, 80
transformation, 67
translation, 83

## U

ultimate reality. *See* Tao
uniformitarianism, 236
universe, 145, 148–49, 155, 160,
   179, 182, 188–90, 214, 234
uracil, 80, 91, 93, 95
uranium 235, 101, 131–32
Urey, Harold, 177
Urey-Miller experiment. *See*
   Miller-Urey experiment

## V

valine, 86
Vatican, 139
viruses, 67
von Baer's law, 118, 120. *See also*
   Baer, Karl Ernst von

## W

Watson, James, 72–73
Watt, James, 100
whales, 107, 126
*When Science Meets Religion*
   (Barbour), 137
whirlpool, 172
Whitehead, Alfred North, 224
Whittington, H. W., 195
Wickramasinghe, N. C., 180
Wilson, Edward O., 142–43, 146

www.ingramcontent.com/pod-product-compliance
Lightning Source LLC
Chambersburg PA
CBHW031831170526
45157CB00001B/260